ENERGY AND AMERICAN VALUES

ENERGY AND AMERICAN VALUES

Ian Barbour
Harvey Brooks
Sanford Lakoff
John Opie

Preface by John Agresto,
National Humanities Center

A Project of the National Humanities Center
Research Triangle Park, North Carolina

PRAEGER

PRAEGER SPECIAL STUDIES • PRAEGER SCIENTIFIC

Library of Congress Cataloging in Publication Data
Main entry under title:

Energy and American values.

"A project of the National Humanities Center, Research Triangle Park, North Carolina."
Includes bibliographical references and index.
1. Energy Policy—United States. 2. United States—Economic policy—1981– 3. Quality of life—United States. 4. Values. I. Barbour, Ian G. II. National Humanities Center (Research Triangle Park, N.C.)
HD9502.U52E4483 1982 333.79'0973 82-13174
ISBN 0-03-062468-1
ISBN 0-03-062469-X (pbk.)

Published in 1982 by Praeger Publishers
CBS Educational and Professional Publishing
a Division of CBS Inc.

23456789 052 987654321

Printed in the United States of America

Contents

Preface

As this book goes to press, the "energy crisis" that began for Americans in 1973–74 has all but officially been declared over. Rising prices have led to sharp decreases in consumption, and the supply of oil is forecast to be adequate for the immediate future, despite international political turmoil. The energy *problem,* however, is by no means behind us. It will remain an acute concern for generations. That continuing problem, especially as it bears on America's social and ethical values, is the subject to which this study is addressed.

Americans and people throughout the industrialized world are critically dependent on abundant supplies of energy. Consider for a moment the effects upon contemporary civilization if all sources of modern energy services were simply to disappear. We would be faced with more than merely the inability to drive our cars, for there would hardly be anything to drive *to.* No factories, few exchanges, scanty markets, empty stores. The urban and suburban character of modern life would be in total disarray, and land and human labor would again become the foundations of all private and social existence. Civilization as we know it would be pushed back three hundred years.

So central is the consumption of energy to our way of life that it is hardly an exaggeration to view the history of energy and energy use as the history of the modern age itself. From brute force and human labor, from wind and water power, through coal and steam, oil, gas, and electricity, the character of modern culture has been shaped by the production and application of primary energy resources. All of the major technological developments that have so changed the nature of life—the locomotive, the cotton gin, the reaper, the sewing machine, the refrigerated boxcar, the automobile, the airplane, radio, television, and the communication satellite—depend on energy sufficient to make them work and work profitably.

In this country such energy-intensive technologies have developed in tandem with many of the most fundamental principles of contemporary civilization—with the belief in liberty, equality, privacy, and material progress. Because we prize the general increase in ease and private leisure, we have welcomed the discovery of a whole array of energy-intensive "labor-saving" devices. Because ours is a culture that puts more of a premium upon the freedom of individuals to move about, live where they will, and go where they choose, rather than on maintaining cohesive or permanent communities or keeping to one's original station in life, transportation, especially private transportation, has been made a central feature of everyday life. Energy and social values have so close a connection that any change in one must shortly have an impact on the other. Therefore, the question of energy supply and use is not merely a technical question to be left only to engineers and power company executives; it is a question that must be of equal interest to historians, social theorists, moral philosophers, and concerned citizens as well.

Until the very recent past, however, there was little recognition that energy could be considered a cultural and moral phenomenon. With usable energy abundant and readily available, the effects on our way of life that would follow from a serious curtailment of relatively low-cost energy were hardly raised. Only since the international oil crisis began to be felt have we begun to make serious inquiries into the place of energy in the world and to see the use of energy as a matter of historical and philosophical importance. Only now have we begun to explore the links between our use of energy and our values. The present anxiety over the supply and price of fuel invites us to ask hard questions about ourselves, our needs, the conditions of our lives, and the needs and rights of others.

For these reasons the officers of the National Humanities Center thought it crucial that scholars outside the technological-scientific community take a hard look at the implications and ramifications of the current energy situation. The aim was not to encourage these scholars to address "the energy crisis" as such but, rather, to review the broader connection between energy and culture, energy and values, energy and liberal democratic ideals. This inquiry was intended to complement efforts being made from the perspectives of technology, policy analysis, and economics—not to substitute for them.

The inquiry begins with the knowledge that questions regarding the impact of energy resources on the quality of life or the relationship of energy development to the history and future of contemporary civilization are appropriate areas for nonscience scholars to examine. For example, we need to have a better understanding of the possible relationship of energy consumption to economic well-being and social harmony. To what degree is a crisis in energy supply likely to become a crisis in social justice, both at home and in the world at large?

We need to understand better the subtle connections between the diffusion of liberty to all and the ready supply of energy and energy-based technology. To what degree has energy abundance contributed to the liberation of the underprivileged, created leisure where there was drudgery, or increased the numbers of those who can leave the demands of the home and hearth to engage in productive endeavors in the marketplace? What, in brief, might energy scarcity mean for the continued advancement of opportunity and living standards? What might it mean for the future of the less developed countries? What choices do we have and how should we consider them?

We need to examine the political implications of energy availability for existing democratic governments and for the future of democracy worldwide. We need to understand more clearly the role that energy plays in supporting economic decentralization and a market economy. To what extent do restrictions on energy supplies or choices or particular energy paths increase the need for greater political centralization? To what degree do all democratic nations presupppose some general level of abundance and ever-increasing prosperity in order to make democracy work smoothly, without civil discontent or factional strife?

We need to achieve a better understanding of nuclear power, its promise and its risks. Technical answers regarding safety and design only go half way, for we need to consider such nonscientific questions as: What degree of risk, both to ourselves and to others, should people be willing to take to obtain the benefits nuclear energy can confer? How can we take into account issues of equity in the distribution of risks and benefits? What obligation does the present generation have to subsequent generations regarding the production and storage of wastes that will remain hazardous for centuries? What safeguards are morally sufficient to prevent nuclear energy from contributing to nuclear proliferation? On the other hand, how should we view the consequences of increased coal burning and the potentially catastrophic results involved in increasing the amounts of CO_2 in the world's atmosphere? Is there a moral calculus that helps us to weigh more intelligently the possibility of major nuclear disaster, however slight, against the constant peril and virtually certain fatalities involved in coal mining?

We need to examine more carefully the political and social consequences of the widespread use of solar and other alternative energy sources in place of conventional supplies. What might the possibility of relying more on decentralized (even individualized) power sources mean for liberty, social cohesion, urban life, and civic interdependence? What are the ethical implications of a rigorous policy of energy conservation? How should we approach the problem of conflicting values, such as when the demands of curbing consumption run up against the demands for work?

Finally, the energy problem threatens to widen still further the gulf between the developed and the underdeveloped nations, the rich and the poor countries of the earth. This gap was far smaller those few centuries ago when all oil was in the ground and electricity only in storm clouds. Energy and energy technology have raised the highly industrialized few far above the level of the preindustrial many, not only in riches but in productivity and in power. It is hardly coincidental that the United States consumes one-third of all the energy resources of the whole world, alone produces from that energy one-third of the world's total of goods and services, and commands far greater military power than all the poorer countries combined. This discrepancy in consumption, production, and power raises serious questions of resource ownership and fair international allocation, questions of the depletion of finite supplies and international equity, and also raises the specter of retaliation and retribution. In this connection, what is just and what is unjust cannot be answered by technical or scientific considerations alone.

This book began with an attempt to focus mainly on America. It was the largely unexamined relationship between energy and our history and values that first prompted this inquiry. Yet it soon became clear that America's future and the future of its principles are connected to the fate of democracy and democratic values in the wider world. In the sphere of energy policy, what affects our allies, our antagonists, and, above all, the developing countries, affects us. America still remains the central focus of this book. But the book was written with a heightened realization that American prospects cannot be understood totally apart from those of the rest of the world.

Here a caution is in order. This is a volume that seeks to identify and reflect on the issues at stake. It cannot hope to give the final answers to all or any of the most troublesome questions. This book would deceive more than it would enlighten were it to promise such solutions. It does intend, however, to take a sober look at the issues, to review the options open to us and the range of answers that have been given, and to offer some guidelines for thinking about problems of this magnitude.

This volume grew out of a study on energy and the values of modern society that took place at the National Humanities Center located in Research Triangle Park, North Carolina. The study began with three Fellows whose areas of interest all fell within the domain of energy and values: Ian Barbour, a physicist and professor of religion and philosophy; Sanford Lakoff, a political theorist and student of science policy; and John Opie, a professor of the history of technology and environmental history. They were soon joined in this project by Harvey Brooks, professor of technology and public policy, recently cochairman of the National Academy of Sciences Committee on Nuclear and Alternate Energy Sources (CONAES), and a trustee of the National Humanities Center.

Although it is a collaborative effort on the part of the four authors, each person was primarily responsible for certain chapters and each chapter still carries the mark and insights of its principal author: Opie wrote the first and second chapters, Barbour the fourth, fifth, and most of the tenth, Brooks the sixth and seventh, and Lakoff the third, eighth, and ninth. But all the chapters were extensively discussed and rewritten throughout the year, and the book is offered as one work, the combined effort of four scholars who shared their ideas and worked closely together. Even though the project started with considerable divergence among the four authors—and some differences in emphasis among the chapters are still evident—nevertheless a working consensus was achieved, not only about the underlying issues, but about the directions in which national policy should move if it is to be consistent with basic American values as well as technical, economic, and environmental constraints.

Over the course of their discussions the authors had the benefit of meeting with a number of people who willingly offered views that helped contribute to this book. Kenneth Boulding of the University of Colorado was part of the original planning committee, and John Voss of the American Academy of Arts and Sciences not only advised this project from its inception but also helped carry it through to completion. Others both at the Center and elsewhere, including James Schultz of the Heller School at Brandeis University, Charles Wolf of Social Impact Assessment, Dorothy Nelkin of Cornell University, Bruce Davie of the Ways and Means Committee of the United States House of Representatives, and John Clark of the University of Kansas and a Fellow of the Center, responded to inquiries and offered advice.

For their patience and perseverance, the authors are keenly indebted to the Center's Librarians, Alan Tuttle and Rebecca Sutton, and to its manuscript typing staff, Karen Carroll, Madeline Moyer, Jan Paxton, Ineke Hutchison, and Cynthia Wilson.

The National Humanities Center and the authors especially wish to thank Thomas Hughes of the University of Pennsylvania, Sam McCracken of Boston University, and Douglas MacLean of the Center for Philosophy and Public Policy at the University of Maryland for their insights and generosity. These three scholars read the original text and together constituted a two-day working conference that devoted itself to a very thorough and constructive review of the manuscript.

The Center would also like to express its thanks to a large number of supporters whose generosity made this study possible. Among these are Westinghouse Corporation, the Committee for Energy Awareness, and the National Endowment for the Humanities. No agency outside the Center had any hand in the selection of the Fellows engaged in the project, on their decision to collaborate, or on the contents or results of the study. The support these contributors gave to the Center in this project, however, coupled with their respect for its scholarly independence, helped make this inquiry possible.

This, then, is a book about history and values. It is not a book of facts and figures, though facts and figures appear in it in sure abundance. It is not a book that explains the details of energy technologies presently working or under consideration; for this readers must look to other sources, many of which are cited. It is, rather, a study of the social, political, and ethical issues raised by those facts, figures, and technologies, and about the varieties of possible and appropriate responses to the issues. The aim is to reflect on these larger issues that surround the energy problem, reviewing the various sides of the debate as fairly as possible. Our hope is that as these issues come to be debated and as policies are made, this study will help make it possible for Americans to find their way more easily through the complexities that the energy problem poses.

John Agresto, Projects Director
National Humanities Center

ENERGY AND AMERICAN VALUES

1 Energy and the Rise of American Industrial Society

THE ENERGY CONNECTION

"Progress" was the official theme of the famed Great Exhibition in London in 1851.[1] The Exhibition centered on the magnificent Crystal Palace, where, according to the fair's sponsor, Prince Albert, the world's greatest industrial achievements were displayed. The hall of machinery reverberated with the deafening noise of locomotives, marine engines, hydraulic presses, power looms, and an Applegate and Cowper printing press turing out 5,000 copies of the *Illustrated London News* each hour. A model of Watt's 1785 steam engine—one cylinder, 40 hp—was placed alongside a modern marine engine of four cylinders, 700 hp. Symbolically, a single huge block of coal, weighing 24 tons, sat majestically in the hall.

More than labor-saving machinery, powered by steam engines, was being exhibited. Also on display was the ideal of progress—the belief that thanks to industrialization, humanity finally had gotten the upper hand over raw nature. To anyone strolling through the rows of machines, it seemed as if, at long last, human history would be the story of man's inevitable advance.

Until just 50 years earlier, in 1800, most Europeans had lived in a rural world of farms and rutted dirt roads, closer to the Middle Ages than to the present, and not far advanced from the emergence of civilization 5,000 years before. But by mid-century, Britain could celebrate a way of life literally unimagined by a prior generation: the Industrial Revolution created a world of iron, coal, steam, machinery and engines, railroads, steamships, and telegraph lines, even ingenious new devices for making life easier, like the indoor water closet, the fixed bathtub with running water, and the gas cooking range.[2] To Albert and most others, technological achievement also carried with it aesthetic achievement—as evident in the Crystal Palace itself—and human ethical progress as well. The whole world had made great advances since the Industrial Revolution began, and the future held even greater promise.

Energy and the Coming of Industrialization

The Industrial Revolution in England that inspired this optimism was the result of an early eighteenth-century energy crisis, as much as of any other single factor.[3] England's forests had been badly depleted in order to make the charcoal that was used to produce iron. The amount of iron produced was negligible measured by later industrial standards, but it was critical to the economy of preindustrial England. In 1709 Abraham Darby developed an early "technological fix" for the problem posed by the exhaustion of wood when he succeeded in substituting coke for charcoal in his blast furnace at Coalbrookdale in Shropshire. Coal had long been available, and its qualities were well known, but the mining of coal was judged to be too dangerous, handling it was dirty and cumbersome, and to make matters worse, burning it fouled the air. All these objections were overcome, however, because of the urgent need to find a cheap and plentiful substitute for scarce wood in the production of iron—a need that coal in the form of coke served, thanks to Darby's new way of using it.

At first, the introduction of coke proved to be beneficial but not revolutionary in its effects. England's forests were allowed to regenerate, an abundant mineral was put to important use, and a continuing supply of iron was guaranteed. Before long, however, this simple substitution of one energy source for another was to transform western and world civilization. Without an abundance of relatively inexpensive and readily available energy industrialization would not have succeeded. These features often carried more weight than other factors, such as pollution, working conditions, or a changing social order.

Energy was only a part of the picture. The growth and spread of industrial production resulted from the interplay of a host of reenforcing elements, political, economic, and cultural, as well as energy resources and technological innovation. In turn, the Industrial Revolution transformed these political, economic, and cultural forces and helped point western civilization on a unique course: a remarkable preoccupation with material growth as an answer to human problems. As the Great Exhibition proudly proclaimed, the unprecedented creation of material abundance instead of humanity's perennial scarcity was a demonstration of the superiority of industrialization as a means to improve the human condition.[4]

Energy had of course always been essential to human survival and to the advance of civilization. The ability of early people to master fire made them aware of their superiority over other animals and inspired them to imagine themselves competitors of the gods, as in the myth of Prometheus. But it was human muscle power, supplemented by that of domesticated animals, that provided most of the energy used throughout the long early history of the human race. At about 2500 B.C. the discovery of the wheel made the use of this muscle energy much more efficient and accelerated the growth of civilizations.[5] Yet, even with fire, work animals, and the wheel, constant scarcity was still a constant and a peril.

In the ancient world, human muscle power was often organized in the form of slavery. Slavery allowed the ancient Romans to neglect the newly-available waterwheel for four centuries—an attitude that prevented them from taking still better advantage of human labor power and from realizing the greater potential of a freer and more cooperative society.[6] The first "power revolution" occurred during the Middle Ages and remained dominant into early modern times.[7] This was the development and widespread acceptance of the vertical waterwheel. This wheel did the work, it was said, of one hundred slaves. Soon there was one waterwheel for every 400 people in England. The windmill was another great labor-saving device: like the waterwheel it could do the work of a hundred men. The horsecollar also appeared in the Middle Ages and was another step toward what we today call "energy efficiency." The combination of horsecollar, horseshoes, and stirrup allowed a horse to do the work of ten men.[8]

For most of human history, fossil and organic energy sources (wood, peat, coal) were used only directly for heating, cooking, and lighting, or for firing the kilns and furnaces used in smelting ores. As a prelude to the Industrial Revolution, however, it came to be recognized that fossil energy, primarily coal for the steam engine, could be used to power *machines,* which could replace hard and tedious work.[9] Such machines could supplement, at long last, the limited endurance of human and animal muscle. Mechanization was also desirable, despite its possible faults, because it would not be subject to the vagaries and seasonal changes of water and wind power. It is hard now to appreciate how this quest to establish regularity in work consumed society's attention; the intermittent and undependable quality of production had made steady progress unattainable in preindustrial times.

Steam power marked a large advance in the effort to make energy available to do civilization's work. Thanks to the introduction of the steam engine, the link between fossil energy sources and the process of industrialization became permanently established.[10] Technological improvements made the steam engine more efficient. James Watt's 1760 version used less than half the coal for the same output as the classic 1712 Newcomen "atmospheric engine," which itself was far superior to Savery's 1698 "miner's friend." Even though it was first thought to be useful only for pumping water out of the increasingly deep coal mines, the steam engine gained widespread use as a cheap, decentralized source of power, and eventually became a decisive factor in general economic progress.

From the eighteenth century on, steam multiplied the strength, versatility, and scale of available power, far outstripping all other energy sources put together.[11] Watt's introduction of rotary steam engine action powered new technologies for spinning and weaving cotton. Textile factories transformed England from an agrarian to an industrialized nation. In the nineteenth century, steam became the motive power for railroads and shipping and began to spin turbines invented to generate electricity. Today the large generating station, whether fired by coal or nuclear energy, supplies all the electricity needed by hundreds of thousands of consumers. The modern power station is a far cry from Watt's simple engine, but the principle remains the same: boiling water to produce steam.

Preindustrial America: Laying the Foundations

Americans had their own Great Exhibition, the Centennial Fair, which opened on May 10, 1876, in Philadelphia.[12] It was criticized for its artistic mediocrity and blatant materialism, but the primary theme of the fair was nothing so much as the triumph of material progress over everything else. The centerpiece was George Corliss's huge steam engine, the largest in the world. When President U.S. Grant, together with the visiting Emperor of Brazil, arrived to inaugurate the fair, they first visited the 30 foot high, 700 ton, 1,400 horsepower engine. It was to produce power for all the machinery exhibits throughout the fair. The president and emperor jointly started the engine by turning handles that built up steam pressure. They then stood back in awe. Public opinion compared the Corliss engine to a great heart whose steady pulsations kept the entire exhibit running. Corliss also made a further advance beyond the sheer scale of the engine; it could regulate its own speed and its own consumption of fuel, depending on the load it carried. What Corliss had accomplished was a long-sought dream: a steady, uninterrupted flow of energy. Like the Crystal Palace Exhibition, the Centennial Fair demonstrated more than materialism. For Americans, the Fair also symbolized the progress achieved since the Revolution by hard work and technological innovation.

It is incorrect to suppose that for much of American history there was no such thing as an energy problem. While industrialization was taking hold in England, the new United States was still a rude "undeveloped" country. Henry Adams, writing his famous history almost a century later, looked upon conditions in the United States of 1800 as unbearably primitive. He noted that the means used to sustain life were not appreciably advanced over those used by the Jutes and Angles in the fifth century A.D.: "Neither their houses, their clothes, their food and drink, their agricultural tools and methods, their stock, nor their habits were so greatly altered or improved.... In this respect America was backward."[13] More recently, the historian Curtis Nettles has noted that the hand tools used by a typical American farmer in 1800 "would have been familiar in ancient Babylonia."[14]

But Americans had one great advantage: the American continent was well-endowed and sparsely settled. The first great task the American settlers had to face was the exploration and settlement of a vast national territory, which would become more than six times the size of the original settlements. The scope of American geography cannot be overestimated. Space, not time, has been the controlling dimension of American history: "empty land" gave this society its distinctive scale and pattern of activity.

Whereas the Europeans could draw on a vast labor pool, Americans had to contend with a shortage of labor. This very shortage was one of the impetuses behind the adoption of slavery in the southern settlements; in the North it stimulated invention. But capital for investment was also in short supply, whereas in Europe, capital accumulation had been proceeding over several centuries.[15] Under the circumstances, Americans made the best of their opportunity by taking advantage of the region's great abundance of land—well-watered, the eastern third covered with primeval forest, composed of some of the world's richest virgin soil, and all in a zone of temperate climate. Without major land and capital resources, Americans had little choice but to concentrate on exploiting easily available resources like land and wood.

Early "manufacturing" depended largely upon the physical muscle and simple equipment of blacksmiths, coopers, masons, carpenters, and weavers. Capital-intensive "factories" consisted mostly of lumber mills, shipbuilding yards, iron-making foundries, glassworks, fishing organizations, and occupations associated with building construction. Traditional waterwheels and the occasional windmill provided the only power available to grind corn and other grains into flour, other than human or animal muscle. Waterwheels and windmills also served to saw wood and to power bellows for forges and furnaces. As late as 1850 wood supplied more than 90 percent of the fuel used for energy in America, mostly for heating.[16] The rest came from small supplies of bituminous and anthracite coal. The abundance of forest resources and water power dominated the American energy picture until after the Civil War.

From the vantage point of a prosperous America in 1890, Adams wondered how the founding fathers must have set about achieving the progress towards the better life they had promised. Was Jefferson's guarantee of "life, liberty, and the pursuit of happiness" to be understood as a constantly improving standard of living? "Life" surely meant more than physical survival alone. "Liberty" required the reduction of material scarcity. "Pursuit of happiness," growing out of John Locke's original term "property," was unlikely without material abundance. As Jefferson himself insisted, a large part of human misery was the oppression induced by poverty.[17] In the new age of humanitarian awareness, it was said, thoughtful people could not be indifferent to human want.

Yet, in 1800 virtually all of mankind lived in want. This had been humanity's historic condition; it remains true that most of the people who have ever lived have been poor and hungry. But by 1800 American society was entering an era, in the words of the modern philosopher, Alfred North Whitehead, "when even wise men hoped." Reflecting on the buoyant optimism of the period, the modern historian Russel Blain Nye writes:

> The fertility of the American soil, the enormous variety and fecundity of its plant and animal life, the vast reaches and unlimited resources of its waters and woods, the salubrity and balance of its climate, all made the United States a mighty stage for new great developments in human history.... The United States, in Jonathan Trumbull's view, was "a land of health and plenty, formed for independency, and happily adapted to the genius of the people to whom it was to be given for possession."... The United States, for the first time in history, gave men the opportunity to put into practice basic principles of society and government impossible to test elsewhere.[18]

Jefferson, Franklin, and other leaders were intensely aware that the new society they proposed must be grounded in material improvement as well as in political liberty.[19] Congress was given express authority "to lay and collect taxes, duties, imposts, and excises, to pay the debts, and provide for the common defence and general welfare." The Declaration of Independence, the Revolution itself, and the transformation of the Articles of Confederation into the Constitution, were in large part responses to economic grievances. By 1800 the test was to see to it that these vague promises of a better life were put into practice.

A pragmatic "American Enlightenment" provided the dominant climate of opinion during the formative years of the new nation. This was an intellectual consensus that affirmed the powers of human reason, the utility of experience, and the essential morality of human progress.[20] Pain, suffering, poverty, and deprivation were not accepted as man's eternal lot but as wrongs that required alleviation. The imperative of the new age was to make ordinary daily life good, rational, and orderly. Notions of "standard of living" and "quality of life" began to receive significant attention in this era.

America's Future Course: Agriculture or Industry?

The proper future material growth of the new nation became the central issue of a major debate between Jeffersonians and Hamiltonians during the late eighteenth and early nineteenth centuries. Which would provide more adequately for the physical and moral well-being of Americans, agriculture or manufacturing? One historian, Marvin Fisher, has called the debate "a morality play, not an economic dispute"[21] because the question was whether agriculture or industry would satisfy Enlightenment ideals of human progress and cultural growth.

At first, most Americans would have echoed Richard Wells's statement in 1774 that "the genius of America is agriculture, and for ages to come must continue so." Crevecoeur's indulgent vision of the American yeoman in his pre-Revolutionary *Letters from an American Farmer* cleverly blended political freedom and economic prosperity with agriculture and morality. The self-sufficient, highly individualistic farmer was America's exemplary model: "The world is gradually settled... the howling swamp is converted into a pleasing meadow, the rough ridge into a fine field... hear the cheerful whistling, the rural song, where there was no sound heard before, save the yell of the savage, the screech of the owl, or the hissing of the snake."[22]

In contrast, the life of a worker in England's new factories was thought to be corrupting and debasing, inherently harmful to body and soul. Looking at the environs of Boston, the traveller John Dix cautioned, "COTTON MILLS! In England the very words are synonymous with misery, disease, destitution, squalor, profligacy, and crime! The buildings themselves are huge edifices which loom like gigantic shadows in a smoky, dense atmosphere. Around them are wretched houses, and places of infamous resort; and blasphemies and curses are the common language of those who frequent them."[23] After all, did not modern coal miners and iron puddlers quickly become infirm and die young? In ancient Greece and Rome, were not only slaves consigned to work the mines and forges because of the inherent physical danger and spiritual risk?

Mindful of the contrast between bucolic farming and industrial squalor, Jefferson in his 1787 classic, *Notes on the State of Virginia,* wrote the definitive American plea for the agricultural tradition:

> Those who labour in the earth are the chosen people of God... whose breasts he had made his peculiar deposit for substantial and genuine virtue.... Corruption of morals in the mass of cultivators is a phenomenon of which no age nor nation has furnished an example.... While we have land to labour then, let us never wish to see our citizens occupied at a work-bench, or twirling a distaff. Carpenters, masons, smiths, are wanting in husbandry: but, for the general operations of manufacture, let our work-shops remain in Europe.[24]

America's strength rested in the independent, privately-owned, small family farm. For Jefferson, the immensity of the land, especially after his Louisiana Purchase of 1803, guaranteed the availability of low cost, abundant farmland as an American birthright and as a haven of safety.

Despite the success of British industrialization, and American envy over its products, advocates of American manufacturing found themselves on the defensive. They were few and gained attention only slowly in the face of the euphoria over the bounty and virtue of American agriculture. However, the Revolutionary War had already demonstrated the strategic role of manufacturing and the importance of national self-sufficiency. In 1787 Tench Coxe pleaded for immediate industrialization:

> It will consume our native productions now increasing to super-abundance—it will improve our agriculture and teach us to explore the fossil and vegetable kingdoms, into which few researches have heretofore been made—it will accelerate the improvement of our internal navigation and bring into action the dormant powers of nature and the elements—it will lead us once more into the paths of virtue by restoring frugality and industry, those potent antidotes to the vices of mankind, and will give us real independence by rescuing us from the tyranny of foreign fashions, and the destructive torrent of luxury.[25]

Advocates of industrialization like Coxe were criticized so harshly that they felt compelled to argue that workers instilled with factory discipline also acquired high moral virtue. In self-defense Samuel Slater added a Sunday School for the children who worked in his textile mill.

Most influential, after some delay, was Alexander Hamilton's 1791 *Report on Manufactures,* which has since made him a prophet among advocates of industrialization:

> The employment of machinery...is an artificial force brought in aid of the natural force of man; and, to all the purposes of labor, is an increase of hands, an accession of strength, unencumbered too by the expense of maintaining the laborer...manufacturing pursuits are susceptible, in a greater degree, of the application of machinery, than those of agriculture.[26]

Hamilton thought human labor devoted to manufacturing was superior to agricultural work: it was "constant" not "seasonal," "uniform" not "careless," and especially "more ingenious" and hence "more productive." Manufacturing, side by side with agriculture, was therefore essential to the security and prosperity of the nation and its people. Only continued surpluses and sustained growth could guarantee material well-being and the development of a "populous, thriving, and powerful nation."

Still, Americans did not rush to industrialize. The pressing need, most Americans believed, was to settle the continent. Americans were also enormously pleased with their unprecedented agricultural bounty; Jefferson appeared to have won the fight. Advocates of manufacturing had to contend with the most successful agricultural enterprise yet known in the history of civilization.

Energy in Agricultural America: Early Growth and its Limits

The first improvements in American well-being did not come through energy systems or industrial technology. They took place in the rich farmlands of southeastern Pennsylvania, the Great (Shenandoah) Valley of Virginia, and soon the newly-acquired (1803) deep black soils of the Midwest. The first major symbol of American abundance was not mechanized energy but food.

As early as 1784 James Fenimore Cooper took plentiful food for granted as he described "American poverty": "As for bread [said the mother in a story] I could take that for nothing. We always have bread and potatoes enough; but I hold a family to be in a desperate way when the mother can see the bottom of the pork-barrel."[27]

Because land was available and conditions were good, the population rose at a staggering pace from four million in 1790 to nine and a half million in 1820, with only 250,000 immigrants. Ninety percent of the work force (aside from slaves) cleared and worked the land. Less than ten percent could in any way be called "employees" in some form of manufacturing. Early in the nineteenth century a French observer wrote: "There is no settler, however poor, whose family does not take coffee or chocolate for breakfast, and always a little salt meat; at dinner, salt meat, or salt fish, and eggs; at supper again salt meat and coffee."[28] In that day, meat was scarce at peasant's tables and chocolate an unknown luxury. While the diet may have been repetitious, humanity's age-old struggle to overcome the scarcity of food was not an American problem.

Just as Americans had an abundance of land and food, the vast stands of trees covering the eastern part of the continent encouraged Americans to go through an extended "age of wood."[29] There was no scarcity of wood to propel them like the English into an economic crisis and revolution. Wood was also used as a heating fuel in no less profligate a fashion. Though fireplaces were gradually replaced by wood-burning stoves, home and business heating remained inefficient. But wood supplies were abundant and cheap. Per capita lumber use was five times higher than in England and Wales.

The early agrarian affluence created expectations of further improvement. Americans in the early nineteenth century came to believe in progress long before industrial technology and high energy consumption took hold. The possibility of greater wealth and material gain was first associated with agricultural surpluses, "wood to burn," and vast land holdings. This attitude put any move toward industrial development at a disadvantage. The high returns in American farming set a high general standard of living, far superior to those in England and Europe. As early as 1760 one astute English visitor wrote about the opportunities enjoyed by immigrant workers to America: "Men very soon become farmers . . . nothing but a high price will induce men to labour [at manufacturing] at all, and at the same time it presently puts a conclusion to it by enabling them to take a piece of waste land."[30] He noted that day laborers are so scarce, and so attracted to farming, that an employer had to pay half a crown in New England for work that would have cost only a shilling in England.

A South Carolina backwoods farmer was typical: "My farm gave me and my whole family a good living on the produce of it; and left me, one year with another, one hundred fifty silver dollars, for I never spent more than ten dollars a year, which was for salt, nails, and the like. Nothing to wear, eat, or drink, was purchased, as my farm provided all. With this saving, I put money to interest, bought cattle, fatted and sold them, and made great profit."[31] Not buffered from the natural world by technology, the frontier farmer was vulnerable to extreme physical hardship, the dangers of accidents, and the uncontrollable power of circumstances. But he was optimistic. He did not experience the grinding hopeless labor and deep pessimism of his counterparts elsewhere in the world. His goals were attainable. Labor-intensive farming, together with limited markets and lack of capital, did keep pioneers on a subsistence level, but with good health, reasonable skills, and basic agricultural know-how, they could manage quite well on 40 to 80 acres of fertile land. On this "owner-cultivated" farm, surpluses and cash, plus the accrued value of the land, were not diluted. Based on a "rude-plenty" standard of living, they took pride in self-sufficiency and independence, and were convinced that the future promised a good and abundant life. In most agrarian settings, ideas of growth and opportunity were unusual, but the American farmer took them for granted. There were few forces inclining him to rush into the cities to work in the factories.

These prosperous rural Americans, who quickly became the envy of Europe and Britain, did not have at their disposal many economic mechanisms that would later be judged necessary to make the transition to a modern energy-intensive society. As we have seen, the labor force was still far from matching numbers with need, and good farming prevented the rapid growth of an urban working class. Large-scale liquid capital to build industrial plants and a consumer-oriented money economy was beyond the reach even of most prosperous Americans. Unifying networks of transportation and communications, to provide rapid and efficient movement of resources, goods, and people, were urgently needed, but canals, good roads, and railroads were still in their infancy in the early nineteenth century. Markets seemed entrenched on local levels. Production of essential goods was small-scale and largely family-centered. Business organization enabling management of complex operations was still decades away.

Most noticeably, as we have seen, energy resources and energy utilization were virtually unchanged since Roman and medieval times. Farming was labor-intensive, relying on long hours of hard physical labor. While the American farmer received extraordinary results from his efforts, there were severe limits to his productivity. If the technology and agricultural know-how of the early nineteenth century had not been dramatically improved, more than half of all Americans would have had to remain farmers to feed the rest. In 1800 it appeared that American society would continue along traditional lines indefinitely, a classic Malthusian-Ricardian world of fixed resources and a growing population doomed to reach its limits wherever and whenever the frontier ended.

The Jeffersonian commitment to agriculture seemed eminently reasonable. Given the vastness of territory and the fertility of the land, the most likely future for the new society was more of the same. This was not necessarily a bleak prospect. In a continuing nonindustrial, low-energy scenario, Americans would have gradually expanded across the continent, or at least to the arid limits of the hundredth meridian. They would have acquired more of the tools, equipment, techniques, know-how, and transportation that were vintage 1800 and improved them. Ultimately, perhaps in 100 or 200 years, they would have run up against finite limits in land, wood, water, food, and mineral resources. Population growth would have reduced resources per capita, and the Malthusian checks to further expansion would have come into play, either to promote conflict and misery or a recognition of the need to accept the regimen of the stationary state.

If this scenario went unfulfilled, it was because the Industrial Revolution transformed American agriculture.[32] The first steps were aimed at increasing the acreage that could be cultivated by a single farmer. Ironically, the improvements made saved Jefferson's idealized yeoman well into the twentieth century. The old-fashioned wheat harvesting cradle could harvest two acres a day. The Hussey reaper of 1837 increased this number to ten acres a day, and the 1849 McCormick reaper made it possible to cover two acres an hour, or as much as 24 acres a day. The cotton gin gave fateful reenforcement to the slave economy of the South. Midwestern corn production was reshaped in mid-century when the hand hoe was replaced by the mechanical cultivator and in the 1890s with the corn picker. The introduction of barbed wire fencing in the treeless Midwest in the 1880s helped open the West, as did meat refrigeration, river steamboat travel, and railroad transportation.[33] The substitution of the tractor for the horse regularized and increased food production. It also saved land. In 1920 about 90 million acres of farmland were used to feed horses and mules, or one-quarter of all cropland then available in the United States. By 1960 less than 10 million acres were needed, a significant shift of land use from animal feed to human consumption.[34]

Through agriculture Americans experienced an increase of abundance, independence, ease, and security. With the appearance of steam power, machines, efficient production, and the other features of industrialization, these very same goals began to take on different meanings and new dimensions.

INDUSTRIALIZATION: AMERICA TRANSFORMED

The American landscape in the early nineteenth century was mainly characterized by small rural villages, individual farmsteads surrounded by cultivated fields, dirt highways, water-powered grain mills, busy rivers, and clean air and water. But by the end of the first half of the century, Americans had invented, borrowed, and occasionally stolen the missing pieces of the jigsaw puzzle of modernization. They pointed proudly to the energy, dynamism, and power of a new industrial landscape—billowing smokestacks of factories, steamboats, and locomotives and crowded canals and city streets—and looked away from the more familiar bucolic setting in which the nation had originated.

Slowly Americans worked their way through the inherent ideological tension between traditional Jeffersonian agrarianism and the full promise of the American Enlightenment and its notions of progress. In fact, the very expanse of the land encouraged Americans to depart from the historic constraints of primitive agrarian simplicity into unprecedented ventures of material growth.[35]

As we have seen, Americans gave special attention to the importance of material improvement by linking it with belief in individual rights, freedom of opportunity, progress, and a superior quality of life. Making a virtue of this egalitarian quest for material gain, Americans quickly adopted a laissez-faire, individualistic capitalism as the economic road to a better future. The result was commitment to the principle of individual ownership over the concern for social or environmental good. When this was coupled with the almost unchallenged priority of scientific progress and technological innovation, most of the now familiar features of American society fell into place. This entrepreneurial zeal, combined with the input of English capital, the discovery of a new labor force (farm girls) and the appearance of another (immigrants), the quick deployment of new and proven technologies, and unprecedented energy use, changed the picture rapidly.

In an 1803 report to the American Philosophical Society in Philadelphia, America's most prominent engineer, Benjamin H. Latrobe, examined the state of mechanized power in the nation, especially steam power. He doubted the potential for steamboats but did note the existence already of five stationary engines, one to pump water for the Manhattan Water Company in New York, another in New York for sawing timber, two in Philadelphia for pumping water and running a rolling and slitting mill, and one in Boston for an unknown manufacturing process. All but one were constructed after 1800; Latrobe failed to indicate whether they were old Newcomen engines or the newfangled Watt type. He added that Americans attempted to "improve" on the steam engine by building wooden boilers, which worked for a time, but then "quickly decomposed, and steam-leaks appeared at every bolt-hole."[36]

Henry Adams described other water-powered manufacturing at the opening of the century, notably Eli Terry's Connecticut clockworks, Asa Whittemore's Massachusetts carding machines for wool, Jacob Perkins's important Rhode Island factory capable of turning out 200,000 nails a day (compared to 100 a day handwrought per worker), and vastly-improved machinery in flour mills. Adams noted that these inventions "transmuted the democratic instinct into a practical and tangible shape" and asked rhetorically about their social good: "as they wrought their end and raised the standard of millions, would they not also raise the creative power of those millions to a higher plane?"[37]

One precocious development that aroused curiosity was the late eighteenth century flour milling operation of Oliver Evans of Philadelphia.[38] He devised methods of transmitting water power so that several machines could be run from the same waterwheel. The grain was mechanically conveyed from one part of the mill to another while it was being cleaned, ground, cooled, bolted, and packed into barrels. But there were severe limits on reliability because of wooden parts and coarse tolerances. Nevertheless, the ideas of assembly line work, automation, and even systems functions existed in Evans's mill in very primitive form. Evans had come upon a "high-technology, labor-saving, continuous-flow, energy-intensive, production process." But this experiment in centralization remained an isolated case for generations.

The Demand for More Energy

As the new nation reeled from the War of 1812, it became clear that the United States could not survive as nothing but a pastoral, agrarian society, no matter how prosperous. Henry Clay urged his "American System" on the American people, and Congress under his leadership passed tariff, banking, and "internal improvement" measures that worked to encourage needed capitalization and industrialization.[39] By 1815, even Thomas Jefferson conceded the importance of manufacturing. National leaders were becoming intensely aware of the severe material limitations of the new nation. Rapid industrialization, using the English example of power technologies, appeared to be the only solution for the nation's survival.

Successful industrialization began with the early subterfuges of Samuel Slater, a young mechanic from England's water-powered textile mills.[40] At a time when the English zealously guarded the secrets of their remarkably successful new powered and mechanized industry, Slater slipped out of the country and settled in Rhode Island. Financed by clothing merchants, he had established in the 1790s the first American spinning mill using water power. He employed only nine persons, all children, but his success led to the appearance of other small mills, eight of which were in profitable operation by 1800. By 1815 the number increased to 94. With these new mills, and the mechanization of interchangeable parts proposed by Eli Whitney, Americans had the foundations of industrialization in sight, but not yet the scale to have significant impact.[41]

Energy use increased at an unprecedented rate in the late eighteenth and early nineteenth century, but lack of capital, labor, and transportation left Americans with "neighborhood manufacturers," small in scale and technologically backward. Up until 1790, the waterwheels of mill and furnace industries were highly inefficient because they were "undershot" (water ran under the wheels rather than by gravity feed overhead). Despite Oliver Evans's innovative mill, power transmission was "so little understood that a separate wheel was generally necessary for each article of machinery."[42] Other factors retarding industrialization included the marked isolation of the American manufacturer, insufficient labor supply, and few sources of capital.

By 1815 the opportunity for future growth was clearly evident, but the ordinary American still lived an agricultural hand-labor existence. The most important need for further industrial development was an increased scale of production. This would lower the unit cost of labor and capital while increasing efficiency and productivity. Again these developments first took place in the New England textile mills. What came to be known as "the Lowell System" solved a major American dilemma not by inventing a new energy source but by "inventing" a work force that was available and reliable.[43] In 1810 Francis Cabot Lowell, while on a visit in England, memorized enough of the equipment and layout of water-powered textile mills to be able to build American mill equipment without British permission. His Boston Manufacturing Company, founded in 1813, quickly dominated the American textile industry. Lowell also managed to avoid the dismal English mill town living and working conditions that had burdened England. He attracted intelligent young farm girls "of good character" by building dormitories and rooming houses, staffing them with respectable housemistresses, and providing libraries and amusements. Towns like Lawrence and Lowell were populated by the new breed of American worker, marked by a spirit of earnestness, moral righteousness, and the aim of financial self-improvement. The French traveller, Michel Chevalier, wrote that "Lowell is not amusing, but Lowell is clean, decent, peaceful, wise."[44]

The textile machine, not powered by hand or animal muscle but by the waterwheel, captured the American imagination. It was apparent that there would be vast opportunities for unparalleled growth, if abundant energy resources currently going to waste could be linked with the new forms of mechanization. The regularity, efficiency, power, and productivity of machines run by water and later steam had not yet transformed American life as they would after the Civil War, but the momentum of change was obvious. The rate of capital growth between 1800 and 1860 rose more than ten times, faster than growth in the labor force, whose work was now measured in "machine-hours."[45] In the same era, productivity increased 50 percent as the labor force began its lengthy shift from agriculture to manufacturing, and as workmen were provided machines that were more efficient. The Gross National Product in 1860 was not only larger but its composition was also quite different from what it had been earlier.

The Jeffersonian goal of an enlightened and benevolent citizenry enjoying a higher standard of living appeared within reach; but America's material well-being was not to be guaranteed only by the remarkable self-perpetuating industrial mechanism Hamilton had imagined. The combination of mass production, interchangeable parts, and tariff protection—Clay's "American System"—was intended to lighten the human burden of physical drudgery while also conquering scarcity. In the United States, this goal demanded higher use of energy than in other emerging industrial nations because of the need to use a minimal labor force as effectively as possible. Energy efficiency was of no great concern, however, because primary energy was cheaper and more readily available than either labor or capital.

Thanks to the availability of this relatively cheap supply of energy, automatic machines were producing nails in Massachusetts and brass pins in Connecticut, simple items which previously had been laboriously made by hand. Soon, the American System gained the grudging respect of the English as it systematically turned out mass-produced complex devices like clocks (1815) and sewing machines (1846). At the British Crystal Palace Exposition in 1851, American technology was displayed in the form of farm machinery and weapons with interchangeable parts.[46]

Steam Power Guarantees the American System

At first steam power was used to replace existing power sources, such as waterwheels and horses, which were less reliable and regular. Such substitution was rarely reversed. The deeper mine made possible by a steam-powered water pump could no longer function under horse power. Larger steam engines combined and centralized factory operations, which would be too expensive and cumbersome to disperse again. High pressure engines in the early nineteenth century made steam equipment lighter and smaller, allowing development of the steamboat, and soon the railroad locomotive. River steamboats were considered so superior to canal, sail, or oared boats that their risks were tolerated as a necessary evil. For the first 25 years after the steamboat's appearance in 1816, 130 boiler explosions took an appalling loss of life. By 1850 probably 1,000 steamboats were lost, with about 3,300 people killed. A public outcry led to the first attempts at government regulation of technology.[47] But the steamboat was far too important to be severely regulated or be forced out of service.

The ability of steamboats to move upstream on America's waterways, which carried the great proportion of trade, made them an early technological marvel. Steam power promised to forge a major series of links between the commercial and industrial East and the agricultural and frontier West.[48] Fulton's *Clermont* ran upstream from New York to Albany in 1807, a remarkable 150 miles in 32 hours. Henry Shreve's use of a flatboat hull and a lighter, less complex, high-pressure steam engine, overcame the difficulties Fulton's heavier deep-draft boats would have had on the snag-ridden, sandbar-filled Ohio, Mississippi, or Missouri rivers. Eastern steamboats were largely used for passenger travel, but in the Midwest carried not only people but vast amounts of goods. By 1860 over 700 steamboats plied the rivers between Pittsburgh, Cincinnati, Louisville, St. Louis, and New Orleans, linking key American regions more effectively than ever before.

Partly because of the need to supplement scarce physical manpower, partly to overcome the vastness of American space, partly to fulfill the material foundations for enlightened progress, and partly to satisfy the dynamic aspect of an American personality, Americans saw in the prodigious use of energy, especially steam energy, the means to material growth. A mechanized society, with seemingly infinite sources of energy, symbolized permanent and progressive human progress.

More than the steamboat, however, the railroad steam locomotive most embodied America's new fascination with energy. It was the modern machine age incarnate, and it remained the prototype of power and dynamism in an energy-intensive society for decades. Following the English lead of the 1820s, and after trials using horse-drawn cars, the Baltimore and Ohio Railroad Company in the 1830s instituted passenger and freight service to Cumberland, Maryland and the West. Soon railroad lines connected all the eastern states, and by 1860 nearly 32,000 miles of railroad had been laid east of the Mississippi. As the historian Charles Moraze put it, the railroads became the "circulatory system of the middle class."[49]

Some historians of technology have described the steam locomotive, together with the railroad system, as the single most significant technological development of any sort in American history.[50] The steam locomotive consumed America's vast forests for wood for ties and fuel. It appropriated the new iron industry for engines, rolling stock, and rails. Perhaps most important of all, it absorbed and rewarded large scale capital investment. The steam locomotive prevented the United States from becoming "balkanized" into independent states or regions. The railroads were a primary motive force for westward expansion into undeveloped frontier regions by making cheap abundant land more accessible. Hence railroad transport also led to a major expansion of American agriculture, as markets could be connected with supplies. The North's railway system helped it win the Civil War. And an early major coalition between business and government was created by the cooperation—critics would say collusion—between westward expanding railroad systems, like the Rock Island Line, and state and federal governments, based on vast land-grant subsidies, rate regulation, and other allowances.[51] Just as lithographs of cities like Pittsburgh depicted industrial might in the form of dozens of smokestacks pouring out black smoke, the steam locomotive was "an instrument of power, speed, noise, fire, iron, smoke...a testament to the will of man rising over natural obstacles."[52]

But while there were only 350 locomotives in use in America in 1838, there were 1,860 stationary steam engines powering rapidly-expanding factories. The flexibility they offered was considered remarkable, since now a manufacturing plant could be located where it would be convenient to markets, labor, and resources, rather than only where water power was available. Factory size could seemingly be increased indefinitely; there was no natural limit to available power.

This transformation produced by the American System has been called the beginning of a "democracy of things."[53] The yardstick of a superior standard of living included not only basic necessities, but increasingly items that made life convenient, comfortable, and "progressive." Items unimagined in 1800, or extremely expensive in 1815, were soon taken for granted as the rightful possessions of a large middle class. Bent pieces of iron were replaced by safety pins, wax paper was superseded by large cheap panes of window glass. The traditional flint and steel fire starter was replaced by the newfangled safety match. Machinery now turned out cotton textiles, carpeting, shoes, "patent" furniture, and tableware; wallpaper became the style instead of paint or leather wall covering. To the list must be added cast-iron stoves, spring mattresses, flush toilets, gaslights, silver-plated tableware, and even roller-shades for windows. Americans of all classes came to believe they were entitled to these benefits produced by machines run by steam and water, and they wanted more.

Coal and Steel Make the Revolution Permanent

Although Americans before the Civil War lived largely in a world of wooden artifacts, materials made of iron gradually became commonplace.[54] Iron textile machinery was far more durable and reliable than wooden machines. Iron turbines improved upon the age-old waterwheel. The unsuccessful wooden steam engine was quickly replaced by the all-iron steam engine. And more efficient and convenient iron stoves—300,000 were made in 1850 alone—replaced inefficient fireplaces. Farm tools included not only the iron plow but, by the 1820s, various hay rakes, cultivators, and reapers.

Belatedly following the English lead, Americans gradually shifted from charcoal to coke to smelt their iron. Anthracite and bituminous coal quickly replaced wood. The efficiencies in scale, and the convenience of coal, were remarkable. A blast furnace that needed 2,000 to 5,000 acres of timber now could be kept running on a half-acre six-foot seam of coal. Production integration and centralization (akin to Oliver Evans's abortive mill of the 1780s) came to the American scene in 1857 with the construction of the Cambria Iron Works in Johnstown, Pennsylvania.[55] It brought together on one site coke furnaces, puddling furnaces, and rolling mills, creating the ability to transform iron ore into finished rails.

But the brittleness of cast iron meant a continued search for a metal with high ductility, toughness, and ability to withstand stress without distortion.

As late as 1850 high-carbon steel was still scarce and expensive, used largely for cutlery and tools. Not until the high-energy Bessemer converter and large-scale production in the early 1870s at the Bethlehem and Cambria works did low-cost steel become a reality.[56] Andrew Carnegie pioneered the vertical integration of the steel industry, combining his interests in coke and pig iron supplies with blast furnaces and steel-rail mills. Carnegie also invested heavily in the high-energy open hearth process, which increased production further. Steel production rose from 16,000 tons in 1865 to an unbelievable 56 million tons in 1915. Even Carnegie was incredulous:

> To make a ton of steel one and a half tons of iron stone has to be mined, transported by rail a hundred miles to the Lakes, carried by boat hundreds of miles, transferred to cars, transported by rail one hundred and fifty miles to Pittsburgh; one and a half tons of coal must be mined and manufactured into coke and carried fifty-odd miles by rail; and one ton of limestone mined and carried one hundred and fifty miles to Pittsburgh. How then could steel be manufactured and sold without loss at three pounds for two cents? This, I confess, seemed to me incredible.[57]

The combination of cheap energy resources, new processes, industrial centralization, and intensive capitalization made American steel extremely cheap. In 1873 it had been $100 a ton, too expensive for steel rails. In one step, Carnegie's 1875 Edgar Thompson Works in Pittsburgh halved the price to $50 a ton. Ten years later it was $20 a ton and by the turn of the century less than $12 a ton, almost one-tenth the price of 25 years earlier.[58] Steel replaced iron as a universal metal for railroad bridges, girders for buildings, nails, and wire. Multipurpose steel beams, nonexistent in 1875, were produced at the rate of millions of tons in 1900.

Steam power had become a typical feature of the American scene, akin to the family farm. Steam engines and boilers still provided the major motive force in the nation. Far higher pressures, undreamed of before, were achieved with safety. Engine speeds rose to high velocity. After the middle 1880s compound steam engines were used in marine, powerhouse, and factory operations.

Between 1865 and 1915 the United States was literally made over by this new coalition of coal to burn, steel to process, and steam for power. In 1900 the annual value of manufacturing was more than twice that of agricultural products, a pace of growth and change that many countries even today would find hard to match. The United States became the leading industrial power in the world in 1895 and by 1910 its output was twice that of its nearest rival. By 1913 the United States accounted for more than a third of the world's industrial output.[59]

At the beginning of the twentieth century, three-quarters of the total energy consumed in the United States came from coal. This compares with 90 percent wood and only 10 percent coal in 1850. In the meantime total energy use more than quadrupled. Coal production still required hard and dangerous work. In 1900 the average coal miner produced only three tons in a day's work. But already coal mine cutting machines accounted for one-quarter of mine production. Mechanized loading and hauling improved labor efficiency, which rose from four and a half tons per man day in 1925 to six and a third tons by 1945. This rose to 13 and two-thirds tons by 1964 following the introduction in 1948 of the continuous mining machine. And in 1964 newly-developed strip mining techniques accounted for a third of all coal production, at the high rate of 29 and a third tons per man day, or almost ten times the productivity of 1900.[60]

Virtuous Enterprise: The "Business Ethic"

In the first half of the nineteenth century, business interests and the entrepreneurial approach were still secondary to the rural American mainstream. By the last quarter of the century, however, the businessman became a highly visible, admirable, even heroic figure on the stage of American life.[61] Inventors, management pioneers, and merchandising geniuses gained public adulation. A hardheaded quest for profits, expansion, and business success no longer needed moral justification, but became a virtue in itself. Henry Adams, in his classic autobiographical essay on "The Virgin and the Dynamo," contrasted the traditional world of agriculture, handiwork, and classical learning with modern prosperity based on coal, iron, steam, and pragmatic materialism.[62]

Who planned the industrialization of America? And for what purpose? Who established official policies to encourage large-scale steel production, electrification, and urbanization? The answer is that no one did with any deliberation. Yet the United States was literally made over in the 60 years following the Civil War. Howard Mumford Jones called it "the Age of Energy."[63] The revolutionary process, the decisions made, the direction of change were left to the market mechanism. The drive for profit in turn moved the market. This acquisitive spirit led to charges of profiteering and monopolies. Advocates of the market mechanism, however, argued that the natural controls provided by competition were the best protection for consumers. The businessman came to be caricatured as the "robber baron," represented by such successful tycoons as Jay Gould, J.P. Morgan, and Cornelius Vanderbilt. They may well have had the predatory habits of Chinese warlords, but their bold financial manipulations helped provide the investment capital that made America an industrial giant.

A preoccupation with money and success attended all ranks of society, not just the top. As Henry Steele Commager has noted, "The self-made man, not the heir, was the hero."[64] A general philosophy of "rugged individualism," symbolized in the "captains of industry" who were praised for raising

American standards of living, became the operating ethic on all levels of society. "The most wretched had aspirations," Charles Beard observed, "there was a baton in every (worker's) toolkit."[65] Andrew Carnegie called the emerging social ethic the "gospel of wealth."[66] America as the "land of opportunity" meant that hard work would lead to riches. Riches were available because high-energy, high-technology heavy industry was said to be the greatest "wealth machine" ever produced in human history.[67]

The idealization of the robber baron, supported by the supposedly benevolent "gospel of wealth" and by a Darwinian "survival of the fittest" philosophy, could not persist indefinitely.[68] As American industrial output matured in the twentieth century, the individual entrepreneur would be replaced by the cautious corporate manager and bureaucrat, practiced less in the art of risk and more in that of negotiation. The one-man, one-plant enterprise grew into large complex organizations, which dominated their industries, vertically and horizontally. By the 1920s the modern business organization was in place, defining American institutional life as much as any other organization, including government and the churches. As many have noted, the American corporation is something like a private government.[69] The exploitation of natural resources, including primary energy sources, demands vast amounts of capital, and the large corporations, interlocked with banks, are the key centers of investment. A close relationship between energy, resources, and corporate America is inevitable.

A major principle of laissez-faire individualism is that general well-being is best served when individual self-interest is left free to maximize personal gain. Justice is best served by leaving decisions over allocation and reward to the impersonal forces of the marketplace. The role of government is to encourage free contractual interchange and cooperation and to keep its own interference with the self-regulating forces of the market to the minimum necessary for security. In response to the threat of monopoly and oligopoly and to demands for alleviation of social distress, this formula has been greatly modified, but in the nineteenth century it was the lodestone and rationale of economic growth.

Industrialization and the Quality of Life

The early settlers had feared that industrialization would introduce inhuman working and living conditions in the New World, as it had in English mill towns. Hard physical work and dangerous surroundings in the new American factories, together with low wages, long hours, and no benefits, led factory workers to seek redress. Before the Civil War strikes and attempts at unionization were severely, even savagely, repressed. As immigrants changed the labor force, wages below subsistence levels continued into the late nineteenth century, working conditions became more dangerous in the steel mills, and the 12-hour day made the condition of the worker harsher than before. Powerless men no longer saw unions as un-American. Not all social improvement came from increased industrial growth; the Knights of Labor and the American Federation of Labor fought to guarantee industrial wages sufficient to sustain a family. A maintenance standard of living was possible for the factory worker only if his wife and children also worked and the family had its own vegetable patch and cow in a tiny tenement backyard. The American farmer once lived in a largely nonmoney economy; soon the worker's life was measured in dollar income.[70] The meaning of the standard of living changed in the process, as did the understanding of the quality of life.

Older worker-owner personal relations were also lost as the size of business enterprise grew. A brass worker in 1883 lamented the new state of affairs: "Well, I remember that fourteen years ago the workmen and the foremen and the boss were all as one happy family; it was just as easy and as free to speak to the boss as anyone else, but now the boss is superior, and the men all go to the foremen; but we would not think of looking the foremen in the face now any more than we would the boss."[71] The nature of work had changed, becoming specialized, fragmented, speeded-up, and less satisfying. It was not done out of a sense of craft but for the wages it earned. Regimented working conditions, with no opportunity for belonging to or identifying with the industrial process, led to dehumanization and labor-capital conflict. The Strike and Panic of 1873, the Railroad Strike of 1885, the Haymarket Affair of 1886, the Homestead Steel Strike of 1892, and the Pullman Strike of 1894 brought on widespread fears of economic collapses, socialist and anarchist "conspiracies," and even revolution.

But workers and farmers did not rise in a sweeping rebellion. The property and lives of the middle class were not threatened. The benefits of industrialization overcame the hardships and the sense of injustice that might have led to revolution. The apparent death throes of American civilization in the 1890s were really its growing pains.[72] Rightly or wrongly, most American workers evidently felt that conditions were improving. The same fragmentation of labor that caused strife also enormously increased productivity. Fewer workers were trapped in physically-demanding tasks. The rise of a highly mechanized industry, which replaced brute strength with steam engines and electric motors, made possible the switch from hard physical work to light-labor mechanized operations.

Productivity, based on mechanical power, increased fourfold between 1865 and 1929. The reallocation of labor, not only from the farms to the factories, but also from heavy industry to consumer goods, and from factory work to service functions, signalled upward mobility to many Americans. The standard of living during the same period rose at a spectacular rate—which helps to explain why America experienced labor strife but no labor revolution. Many Americans enjoyed a transition from working class subsistence goals to middle class aspirations, often regardless of original income or status.[73] Mobility from "blue collar" to "white collar" jobs became part of the American Dream. Most Americans believed they belonged to a middle class and had left a working class life behind them. In 1865 only slightly more than one American in ten was employed in education, the professions, personal services, trade, or government; in 1929 the number was one in five. By 1950, two of every three Americans were employed somewhere else than the factory and farm.

By any external measure, the standard of material life that Americans claimed around 1900 was remarkably high. Even by the best of European standards it was very prosperous. Americans began to demand a plentitude of goods at modern prices, with superior durability, and convenience of use, and which would create new, interesting, and useful activities. No home could be without wood stoves, sewing machines, "Cabinet" furniture, carpets and fabrics, clocks, china, glassware, and the horse-drawn buggy. Yet in 1900, 60 percent of all Americans still lived on the farm, and rural households set the pattern for American consumer expectations well into the twentieth century.

The opportunities created by the universal rush to centralized energy development quickly took hold. The historian Richard M. Abrams has described the United States in the first quarter of the twentieth century as "a land of promise where one person's gain was another person's opportunity, and the inevitable were not just death and taxes but improvement and growth."[74] Even a reformer like John Haynes Holmes found the early years of the century remarkably benevolent: "Those of you who did not live in that period before 1914, or who are not old enough to remember it, cannot imagine the security we enjoyed and the serenity we felt in that old day."[75]

Abrams cautiously concludes that the United States in that era was the most successful society yet in the history of civilization. Anchored more securely than any other society in the material abundance created by mechanization and high energy use, the United States between 1900 and 1915 was perhaps the archetypal industrial society. Material well-being, measured by Gross National Product, seemed to rise constantly and permanently. On the solid foundation of heavy industry came production of consumer goods: "Americans began to reap some of the benefits of their labors." One of the first signs of the coming of a postindustrial era was the greater dollar value of the meat packing industry over the steel industry in 1904, a difference that doubled by 1915.[76] Eating habits were transformed with the development of refrigerator cars, since fresh meat, fruits, and vegetables could be shipped long distances. Buying habits changed with the Parcel Post Act of 1912, and mail order companies came into their own. Leisure time became an important consideration as the workweek declined to 60–55 hours for industrial labor and to 55–50 hours for skilled labor.

The industrial revolution, it seemed, was fulfilling all the promises of the prophets of progress. Especially in America, where resources were abundant and opportunity unfettered, the age-old struggle to overcome scarcity could be waged with confidence that the successes already achieved would not prove only temporary. There was little recognition that the satisfaction of old needs would raise new expectations, or that at some point these expectations would run up against a variety of limits to perpetual economic growth.

2 Energy and Abundance: Advance and Retreat

ELECTRICITY AND AMERICAN LIFE

Between 1865 and 1929 total energy use in the United States rose prodigiously from 160 million horsepower to more than one and a half billion horsepower. One growing segment of this enormous increase belonged to a new development: electricity. A matter of commonplace confusion must be avoided: electricity is not a source of energy. Coal, petroleum, natural gas, water power, nuclear fission, and solar radiation are sources of energy; electricity conveniently carries and distributes energy. When first developed, it was a new way of using energy generated by the flow of water or the burning of fuel. Electricity had so many benefits that it brought on a remarkable change in the use of these primary energy sources. Between 1902 and 1919 Americans increased consumption of electrical power ten times, from 2.5 kwh to 25.4 billion kwh. By 1920 it accounted for 8 percent of all energy consumed in the United States, by 1955, 17 percent, and by 1978, 29 percent.[1] In the first 30 years of this century, investment in utilities, lighting, machinery, and appliances outstripped even the bourgeoning automobile and related industries.

The major advantages of electric power are that it is flexible and divisible. The power plant can be separated from the manufacturing plant, even by long distances. Cumbersome devices for translating and transmitting motion (i.e., from to-and-fro to rotary motion) are unnecessary. Electricity is "on-tap" when required, whether for large or small needs. In time the electricity industry became unique because of its size, its pervasiveness, the variety and diversity of its influence, and its direct social and economic impact.

At first, early in the nineteenth century, electricity was used almost solely for magic and stunts. It received serious public attention only when Samuel F.B. Morse successfully demonstrated telegraphy. Morse's breakthrough came when he developed an adequate electric battery. The first telegraph line was strung between Baltimore and Washington. By 1848 lines went as far as St. Louis. 1861 marked the first transcontinental line and 1866 the first successful Atlantic cable. Ten years later Alexander Graham Bell demonstrated the telephone at the Centennial Exposition in Philadelphia, transmitting the human voice and other sounds over wires. The telephone spread rapidly: by 1887 A T & T had 170,000 subscribers. The first long-distance line connected New York and Chicago in 1892. By 1917 there were more than 11 and a half million telephones in use.[2]

After 1880 interest in incandescent lighting raised a new and growing demand for electricity. Lighting required more powerful electric transmission compared to the mere trickle needed for telegraphy and telephone. Arc lighting was first offered commercially in the 1870s, but it was too bright and too expensive for domestic use. In 1878 Thomas Alva Edison devised a lighting filament that was inexpensive, durable, and would not decompose and blacken the inside of the bulb. He first succeeded with a carbonized sewing thread and bamboo filament. By 1882 the New York Pearl Street Station became the first central power station, distributing its current to 1,400 lamps in 85 buildings. But direct current transmission was limited to a quarter-mile for 110 volts and a half-mile for 220 volts, the latter voltage being a practical limit.[3]

From the beginning Edison perceived his electric lighting not simply as a lamp but as an energy system.[4] He had a vision of a central source of electricity that would light houses and businesses over a large section of a city. Customers could use light as needed; unlike the arc lighting of streets, the failure of one light would not affect the others. Centralization would also allow easy metering of consumption. Edison foresaw the distribution not only of lighting but also of power. At Menlo Park he and his associates worked not only to design the light bulb, but also its convenient screw base and socket; so simple and clever was this patent that it gave his competitors fits as they vainly sought alternatives. Out of the Menlo Park laboratory also came fuses, insulators, regulating devices, conduits, junction boxes, and bigger and better dynamos. The result was the rapid spread of electric power generation; by 1891 there were 1,300 incandescent lighting central stations in the United States, capable of lighting three million lamps.

And there was much more yet to come. Despite Edison's strenuous and wrongheaded opposition, George Westinghouse worked to devise a practical alternating current supply, which would allow energy to be transmitted over long distances.[5] Westinghouse acquired the patents of Nikola Tesla for alternating current and refined them for practical development. In 1893, Westinghouse's banner year, he landed a contract to build dynamos for a major hydroelectric development at Niagara Falls. Niagara-generated power went on sale in 1895 in Buffalo, 25 miles away, and the era of long-distance power transmission began. In the same year the Westinghouse Company won the contract to light the prestigious Chicago Columbian Exposition. Visitors were enormously impressed by the giant dynamos, weighing 75 tons each, and the marble switchboard 1,000 feet square. At night, when the great white buildings were bathed in the light of 5,000 arc lamps and over 100,000 incandescent bulbs, the grounds seemed a wonderland. By 1890, only ten years after Edison's first light bulb, the annual sales of the three big electrical manufacturers—Edison, Thomson-Houston, Westinghouse—were about 25 million dollars.[6]

Until 1900, the generation of electricity was no model of efficiency. Steam engines ran dynamos, which then actually produced the electricity. The widespread use by 1900 of the new steam turbine made the economies of scale of highly centralized electrical generation obvious. Although overshadowed by more glamorous and controversial technologies, the high-pressure steam turbine, invented by Charles A. Parsons in 1884, became America's energy workhorse, as it has been ever since. Factories that had been producing their own electricity found that the large centralized turbines, with higher steam pressures and temperatures, owned by utilities, could produce electricity cheaper. In 1910 a generating plant of 25,000 kw was large. Steam generation of electricity reached new measures of size, strength, and efficiency during World War II as it became the source of power for American naval ships and massive electric dynamos. In the postwar era to the present, steam turbines have generated 83 percent of America's electric power and 75 percent globally. By the early 1960s, generating plants of a million kw were appearing. The simple design of the steam turbine—blades on wheels made to spin by superheated steam—continues to be the primary device for turning raw energy into electricity.

So effective was the steam turbine that it sometimes inspired rhapsodic praise. In 1926 Chester T. Crowell exulted in the *Saturday Evening Post* that the advent of this new mechanism had provided "Nine Slaves for Each Citizen."[7] He compared them to the waterwheel, which created "one slave for each four citizens" in medieval England. He lavished praise on the New York Edison steam turbine at the foot of 14th Street, which consumed 60,000 pounds of coal not in a day but every hour and produced 80,000 horsepower to free ordinary citizens from drudgery. The steam turbine generating plant set up in Cincinnati in 1925 was said to produce the energy equivalent of the work of 9,000,000 men, serving a metropolis of 1,000,000—hence the figure of nine slaves for each citizen. As Crowell noted, when the Cincinnati plant was dedicated, Owen D. Young of New York Edison explained its larger meaning:

> This is the way America must solve her problem of maintaining higher wages than any other country in the world and at the same time keeping her goods competitive in foreign markets. We must put more energy back of the

> worker, in order that he may be a director of power rather than a generator of it.... I want to see this art not only run the giant industries of the cities, but I want it also to be so humble and true in its social service that we shall banish from the farmers' homes the drudgery which in the earlier days killed their wives. We have come here to dedicate a power plant—an instrument of utility. Is it only that? Perhaps it is a temple.

Crowell noted that electric-powered machines brought "a revolution compared with which Marxian Socialism or Russian Bolshevism would be puny and futile, a mere dance in the wind." Life in the new America was a "great, romantic adventure" of "beauty and comfort" for the average person.

Today, such ebullience over energy technologies has a naive ring, but in the early decades of the century it obviously reflected genuine excitement. Crowell concluded that the creators of the machine age, even more than political reformers, prophets, or poets, had liberated the common man: "When mechanical slaves replace human drudgery the general welfare of mankind must inevitably be advanced." Energy and the pursuit of happiness are directly linked: "the time has already come when the variations between the standards of living in two different countries closely approximate the difference between the amounts of mechanical power they use...the ratio between primary power and standards of living will be predicted with almost as great accuracy as are ocean tides." The results of mechanization, he added, are tangible, providing real services to mankind through "the liberation of elbows and the enfranchisement of brains." He coined an appropriate slogan for the era: "Knowledge is horse power and kilowatt hours." Crowell ended his dramatic essay by writing euphorically: "Humanity is just peeping over the mountain tops, catching the first glimpse of dawn in a new world. They have boundless faith that men and women will find this new world lovely."

Lovely or not, growth in electrical generation was certainly spectacular. For approximately the first 50 years of the twentieth century, when total energy consumption grew fivefold, electrical generation grew more than 100 times, or 20 times faster. In the same period, efficiency in the use of fuels also improved. In 1900 it took almost seven pounds of coal, or the equivalent, to produce one kilowatt hour of electricity. In 1920 it took only three pounds and by 1955 less than one pound—a sevenfold increase in efficiency.[8] (Even so, however, more than half of the raw energy is lost in the process of conversion to electricity—a point that modern advocates of thermodynamically appropriate energy use often raise against it.)

Following the use of electricity in communications and lighting, the next major development was its adaptation to machinery.[9] The first uses of electric motors to turn machinery took place on a limited but practical scale in the 1880s. The advantages of electric motors were immediately obvious and quickly made them universal. Exact power ratios were available for each machine. The motor could be cut off and on as necessary. Power could be distributed throughout a plant without the complex, inefficient, and dangerous shafting and belting previously required. Factories became better lighted and more open. Electricity allowed more efficient organization of production. The electric motor mounted on the machine delivered 70 to 90 percent of its power to the machine, while a steam engine, driving a series of machines by the usual system of belts and pulleys, had a 10 percent efficiency. The rapid increase in productivity following the First World War is directly linked to the growth of electrification in manufacturing. In 1901 only 2 percent of all industrial machinery was powered by electricity; by 1929 it was 80 percent.

In 1888 electricity from a central station was first used to power a streetcar system in Richmond, Virginia. This was so superior to horsecars that three years later over 5,000 streetcars were in operation and by 1900 horsecars had disappeared. By 1914 a traveller could go from New York City to Portland, Maine, or to Sheboygan, Wisconsin, on the new, clean, efficient, safe, and reliable interurban electric lines. The first elevated electric railroad appeared in Chicago in 1883 and the first electric passenger elevator in 1892.

Electricity was also responsible for a dramatic change in domestic life. Although appliances did not completely festoon the kitchen until after World War II, early appliances were attached to light fixtures, and soon included electrified fans, hot plates, cigar lighters, stew pots, irons, and sewing machines. By 1900 the wall outlet and plug appeared, together with the heating pad, curling iron, and glue pot. By 1910 the array of electrified gadgetry included electric frying pans, toasters, portable grills, waffle irons, and cornpoppers, together with exotica like electric shaving mugs and sealing wax heaters. By the affluent 1920s they included electric stoves, vacuum cleaners, hair dryers, washing machines, dishwashers, and such novel necessities as marshmallow toasters. Electric thermostats allowed for the first time remote control of heating and later of cooling.[10]

The convenience, efficiency, and flexibility of electric motors made the machine even more pervasive and all-embracing in everyday life. The revolution of convenient mechanization to serve all purposes, some liberating, like washing machines, and some of more trivial utility, changed both home and office life. The Xerox copier is the result not only of optics and chemicals; its development presupposes electrical mechanization. The need for electric power became as universal as that of heating and lighting, and created a lifestyle far more comfortable and convenient than the richest medieval emperor or oriental satrap could boast. Lenin had said that the essence of the Bolshevik Revolution was "the Soviets plus electrification." Lincoln Steffens, who had been an early enthusiast for the new Soviet way, ended up saying of American prosperity in the late 1920s, "Big business in America is producing what the socialists held up as their goal: food, shelter, and clothing for all."[11] Macy's department store in New York advertised, "Goods suitable for millionaires at prices in reach of millions." Perhaps more from electricity than from any other power system came a sense of national well-being: "food, clothing, and shelter for all" became goals within the reach of all Americans, not merely distant dreams.

LIQUID FUELS AND THE AUTOMOBILE

Oil came into American lives because of an early energy scarcity. Even a preindustrial society needs lighting for homes and shops and grease for its wooden and iron gears in mills, canal locks, and axles. Animal and vegetable oils became more scarce and costly as demand increased. Petroleum, which was known from seepages in the ground, occasional oil springs, and salt wells, was an obvious new source of lubrication. Entrepreneurs recognized that a vast market could be developed if capital interests could be persuaded to search for large quantities of oil. The rest of the story is well known. On August 27, 1859, oil was struck at 70 feet at Titusville, Pennsylvania, and the rush for "liquid gold" was on.[12] One year after the first well was drilled, 15 refinery plants had sprung up in the area. By 1863 Pittsburgh had 60 plants with a capacity of 26,000 barrels weekly. Wooden and iron pipelines fed railheads, and soon hundreds of wooden tank cars made oil a widely-available national energy resource. Production rose from 500,000 barrels in 1859 to almost four and a quarter million barrels ten years later. This was equivalent in energy to nearly a million tons of coal. While it had taken the coal industry more than 80 years to develop that energy level, oil production required only ten years to reach it. But in 1900 expert opinion said supplies of oil would last only ten more years. From the first, the oil business was highly speculative and potentially rewarding beyond ordinary expectation. Prices and production fluctuated enormously.

Early oil exploration involved considerable waste. Primitive oil production technologies could not control oil flow. Gushers, pressured by natural gas, blew thousands of barrels daily into the air until they were capped by hard, dangerous work. The first Spindletop well of 1907 blew off and burned as much as 100,000 barrels a day for ten days. Gushers, overproduction, oil and gas seepage, and simple burning off of natural gas as a nuisance, became commonplace. In 1913 the oil field of Cushing, Oklahoma lost more in potential natural gas value than all its oil was worth.[13] Between 1922 and 1934, one and a quarter trillion cubic feet of gas was wasted daily in the oil fields, the equivalent of 250 million tons of coal. Even as late as 1950 more than half of the natural gas was burned off because prices were too low. This was trillions of cubic feet a year of a nonrenewable resource.

Crude oil was first widely used without processing. Gradually, refined petroleum products entered the market, notably kerosene, distillates, lubricating oils, and later gasoline and petrochemicals. About 1900, a 42-gallon drum of crude yielded upon refining 24 gallons of kerosene, six gallons of fuel oil, and five gallons of gasoline. In 1913 the new revolutionary cracking method increased gasoline yield more than 10 gallons per barrel, and yield has increased since.

Energy historians Sam H. Schurr and Bruce C. Netschert distinguish between three eras of petroleum use: "the period of illumination," "the fuel oil period," and "the internal combustion fuel period."[14] Until the first decade of the twentieth century, kerosene accounted for most of oil production. It was used as an illuminating fuel and for heating and cooking. Today it continues as an agricultural equipment fuel and jet aircraft fuel. Distillates were used mostly for heating, steam generation, and diesel fuel.

The third period began with the development of the internal combustion engine. Early on, the most successful automobile engine was not driven by internal combustion but by steam. Electric cars were known at the turn of the century but were expensive and had to be recharged frequently. Internal combustion engines had been known for 200 years, but attempts to make them operate on benzene from coal, illuminating gas, and even gunpowder had failed to make the "explosion" engine successful. It was the marriage between the internal combustion engine and the new fuel, gasoline, that started the revolution of the automobile in the late nineteenth century. Gasoline-powered motor cars were so relatively efficient that the basic principles have not been significantly modified for the last 90 years, a long time for an energy mode.

Gasoline gradually took half of all oil consumption in the United States. Between 1925 and 1955, annual gasoline consumption per motor vehicle rose from 473 gallons to 790 gallons. By the 1970s half of America's energy consumption took place in internal combustion engines. As much as the steam locomotive symbolized the forward thrust of American life in an earlier day, so the automobile reshaped the American way of life by the second quarter of the twentieth century. It extended long-cherished views of mobility, individuality, growth, freedom, and power. But the auto also symbolized American materialism and waste, especially in its fuel consumption and in the energy expended in its manufacture.

The automobile transformed the growing metropolitan regions of the nation. It tied together city and country. It helped destroy America's small towns even as it built up suburbia. In 1900 the average distance to work was one mile; most people walked to work. In the 1970s they commuted by car an average of 15 miles. American lifestyles, from courtship to work habits, from place of residence to use of leisure time, from the experiences of childhood to the practice of religion were all changed by the popular appeal of the automobile. The auto became the second most important investment of the large American middle class, superseded only by the private home. To some Americans the car was their most valuable possession by far.

Any present or future oil shortage would be greatly alleviated if the balance between mass transit and the private automobile could somehow be changed. Such radical change would be felt to be a regressive step, almost a repudiation of the "American way of life." The arguments against the personal auto are strong: it can easily be blamed for raising accident rates and pollution, for expanding highway pork barrels, urban and rural decay, industrial dirt and ugliness, and the irresponsible waste of energy. But the American economy and the American lifestyle can hardly be imagined without it, and without its related manufacturing industries, oil production, road construction, travel-oriented tourism, and unnumbered other aspects.

NUCLEAR ENERGY: CAUGHT BETWEEN TWO WORLDS

On September 24, 1977, Toledo-Edison's Davis-Besse Unit 1 nuclear power plant, located at Oak Harbor, Ohio, between Toledo and Cleveland, operating at 9 percent full power, experienced a malfunction in its cooling system. A pressure-relief valve jammed open and released water that was cooling the reactor. Without this cooling the reactor would overheat and quickly reach meltdown conditions—the infamous "China Syndrome." The main feedwater system failed, and the emergency feedwater system came on, but because of an easily misread water-level gauge, the plant's operators concluded that the reactor was receiving too much cooling, and shut off the emergency cooling pumps. As a result, water poured out of the reactor cooling system and none was allowed in. It took 22 minutes for the operators to discover the open relief valve and to correct the failure before damage could be done.

Eighteen months later, on March 28, 1979, a strikingly similar malfunction took place at Metropolitan Edison Company's Three Mile Island Unit 2, working at 97 percent capacity, ten miles south of Harrisburg, Pennsylvania. In this case, the pressure-relief valve stuck open, the water-level instruments again were improperly read, the emergency cooling pumps were turned off, but it took about two hours and 20 minutes to discover that the relief valve was open. A partial meltdown took place and the plant is still radioactive and closed. While malfunctions in cooling systems of other nuclear plants have been corrected automatically or by plant personnel, at Three Mile Island the first "general emergency" ever to arise in a commercial nuclear power plant was declared.[15]

One set of commentators cites the Three Mile Island accident as evidence that the nuclear industry, the utilities, and the government regulators have been exceptionally careful and successful in developing nuclear energy. It was the first such case, after years of generating nuclear power, no one was injured or killed, and the technology worked exceptionally well: final blame was placed on the operators in the control room, who were held to be inadequately prepared for their responsibilities. Another set of observers point to the same accident as evidence of the dangers of relying on nuclear reactors. In particular they cite inherent design flaws, lax attention to safety problems by the manufacturer, the utility, the operators, and the regulatory agency, as well as the potential for catastrophic harm in the event of an uncontrollable accident. Critics say we were lucky; advocates say they were proven right—the system works.

From the day—December 2, 1942—that the first man-made nuclear reactor reached critical mass, until December 18, 1957, when the Shippingport commercial reactor went on-line, 15 years had passed. In Richard Rhodes's words, "That is rapid development or surprising delay, depending upon one's perspective."[16] Under the Atomic Energy Act of 1946, atomic energy was made a government monopoly; all discoveries and inventions concerning atomic energy were defined as "born" secret. All fissionable materials were the property of the U.S. government. Only by government contract could a reactor be built, and it could not become private property. The Atomic Energy Commission (AEC) was called "the most totalitarian governmental commission in the history of the country."[17]

In 1948, the policy of the Congressional Joint Committee on Atomic Energy was "that reactor development should proceed with all possible speed." In 1949, in an unusual step, the AEC contracted with Westinghouse to begin work on a power reactor for submarines. Hyman Rickover was put in charge, and it would be Rickover who later personally directed the building of the first civilian reactor at Shippingport. The Navy's high-pressure, water-cooled reactor came to become the standard for American industry.[18] In 1953, the AEC concluded that "now is the time to announce a positive policy designed to recognize the development of economic nuclear power as a national object...to promote and encourage free competition and private enterprise."[19] On October 22, 1953, the AEC announced that an AEC-owned nuclear power plant of 60,000 kw would be constructed at Shippingport, to be built jointly with Westinghouse Electric Corporation and the Duquesne Light Company of Pittsburgh. Duquesne Light made its commitment because of Pittsburgh's serious coal pollution problems and AEC's promising economic subsidies and guarantees. President Eisenhower made his famous "Atoms for Peace" speech before the General Assembly of the United Nations on December 8, 1953, Rhodes writes that "'Atoms for Peace' did...encourage Congress, industry, and the press to consider nuclear power, and notably private nuclear power, as peaceful, patriotic, and benevolent."[20]

Before the federal government offered inducements, industry had not rushed forward to embrace the new energy technology, because it was not considered commercially attractive. Conventional electric power generation cost about four to eight mills per kwh; Shippingport first cost 55 to 60 mills per kwh, but the AEC sold it for eight mills to Duquesne Light.

The Atomic Energy Act of 1954 established new patterns to encourage nuclear power. For the first time it allowed private industry to own and operate plants; the government monopoly was broken. Rickover, hard at work at Shippingport, told the Joint Committee, "I think we have babied a lot of people in this country too long with the glamour of atomic energy.... I think as soon as possible we have got to get down to do it like any other business." But he also said, "All we have to have is one good accident in the United States and it might set the whole game back for a generation."[21]

Although given a virtually free hand by government, the nuclear industry still had to compete in the free market, and quickly learned that nuclear energy, in Alvin Weinberg's regrettable statement, was not going to be "so cheap it could not be metered." In the 1950s and 1960s electric rates from conventional systems, like coal-fired power plants, were actually falling. At the same time, Shippingport was experiencing construction cost overruns of 50 percent. The industry quickly learned that nuclear plants could compete only if they were several times larger than conventional plants, thus benefitting from economies of scale. Very large scale plants ran a high risk of overheating more intensely and rapidly during a cooling system malfunction, but this risk had lower priority than the potential cost advantages.

In 1974 Congress established the Nuclear Regulatory Commission (NRC) as a partial successor to the old Atomic Energy Commission. The NRC was authorized by law to develop new regulations covering every aspect of plant design and construction down to the last nut and bolt and last rule in operating handbooks. But the only basic changes that actually resulted involved an effort to streamline the licensing process. To many critics it appeared that comprehensive and strict regulation came a poor second to the continued government commitment to rapid nuclear reactor expansion at the hoped-for rate of "one or two per week for every week in the years from 1975 to 2000." Peter Bradford, one of the NRC's five commissioners, said that the regulatory process fell into "fundamental disarray." The conflict was over two sets of priorities: expansion of nuclear power, without undue and costly regulatory delays, or a more active federal supervision of nuclear safety, with inevitable delays "in the public interest." The Presidential Commission set up to investigate the accident at Three Mile Island severely attacked the NRC for being "unable to fulfill its responsibility for providing an acceptable level of safety for nuclear power plants."[22]

In the wake of Three Mile Island and of the resulting public outcry, the further development of nuclear power has been enmeshed in heated controversy. As one analyst has remarked, the nuclear debate is "proceeding with the intelligence, grace, and charity of a duel in the dark with chain saws."[23]

ENERGY AND ABUNDANCE

In 1950 the historian David Potter examined American affluence in his book *People of Plenty*[24] and concluded that the achievement of this society's declared values depended heavily on the success of its industrial efforts. American industry makes the most of human labor, reduces hazardous or degrading work, exploits natural resources efficiently, and promotes well-being by producing cheap durable goods and services. From this perspective, more industrialization, not less, is the proper course if American values are to be promoted.

Potter noted that mankind has lived with scarcity throughout history, and that American society was the first to break out of the pattern and build a civilization based on abundance. Abundance has both economic and political effects. Economically, the promise of a better life is realized with the availability of "food, clothing, and shelter for all." Politically, abundance reinforces belief in the founding ideals of liberty, equality, opportunity.

By the 1950s, belief in mass consumption seemed to dominate other aspects of American life. The goal was "more, bigger, better, and still more convenient." This consumerism depended on a privileged use of energy and other resources. Over 40 percent of the world's aluminum and a third of the world's fuel is consumed in this country. While it is impossible to measure personal happiness, a sense of individual freedom, fruitful leisure, and other satisfactions, Americans have translated many of these goals into material terms. And these material satisfactions have usually been associated with energy-intensive production.

As Potter showed, economic growth forms part of the promise of democracy. It has encouraged a social climate for continuous "modernization" through scientific discovery, technological innovation, business enterprise, capitalistic enterprise, and successful management of large, complex systems. The belief in quantitative growth has become a traditional conviction.

Efficient use of energy has been one of the most important conditions of material progress. In the 1850s, energy consumption per person was the equivalent of four tons of coal; by 1880 it was down to two and a half tons; in the 1950s to less than two tons.[25] Solomon Fabricant describes this technological improvement as society's "intangible capital": "all the improvements in basic science, technology, business administration, and education and training, that aid in production."[26] Energy efficiency was 8 percent in 1850, 12 percent in 1900, and over 30 percent in 1950. A diesel railroad engine did the same work as a coal-fired steam locomotive at one-sixth the energy consumption. The long-term trend was away from direct consumption of raw energy to the use of processed and converted energy products, as in the change from coal to diesel oil for the railroads, or the nationwide switch to electric power generation.

In the 1880s a little over 26 million horsepower was consumed in the United States, slightly more than half a horsepower per person. Almost half of this was from work animals and less than half was from fossil fuels. In the next hundred years the use of fossil fuels multiplied four times (as did the population). But with higher technological efficiency, by 1970 horsepower per capita was over 90, or almost a 200-fold increase since 1880.[27]

Abundance, and the democratic liberties it produces, also rests upon a "production ethic." Fundamental responsibility rests with private initiative aided by government. From this perspective, America's greatest problem is the troubled state of its industrial capacity, especially the decline in the rates of innovation and productivity.

But the assumption that growth in industrial production is a linear function of energy consumption needs reexamination. Total energy consumption over the last hundred years has increased at less than half the rate of the Gross National Product.[28] There is good reason to suppose that a relatively high rate of economic growth can be accomplished without a corresponding increase in energy consumption—provided capital and technological ingenuity can once again come to the rescue.

HISTORIC CRITIQUE OF ENERGY AND INDUSTRIALIZATION

Casting doubt on the benefits of the new industrial system has been a constant contrapuntal theme during its history.[29] Criticism of the capitalistic merchant and iron-working guild "mechanick" is actually preindustrial, since they were seen as latter-day intruders upon the feudal order of society and were never allocated a place as links in the all-important sacred-secular "Great Chain of Being" that controlled all medieval relationships. With the coming of the Industrial Revolution, the semimythical Luddites between 1811 and 1816 attacked English textile mills and smashed power looms. They blamed unemployment and low wages on the machines. English farm workers attacked the new threshing machines in 1830. French textile workers used their wooden shoes, *sabots,* from which the word *sabotage* comes, to kick machines to pieces. Complaints had long been lodged, particularly in England, and laws passed, against the pollution and working conditions of mill towns.

Critics also attacked the mechanization of life which the new steam power imposed. In contrast with the craftsman's pride in his handiwork, skill was instead built into the machine. The new factory organization of time and the division of work into separate disconnected operations were designed to make the most of machinery but turned out to be detrimental to job satisfaction and well-being. "It was forgotten," Bertrand de Jouvenal wrote, "that 'hands' also had hearts and minds as well." The loss of time-honored principles of a "fair and just wage," the growth of an impersonal relationship between owner and worker, and the single-minded obsession with "productivity for profit" seemed too high a price to pay. "The new organization of work by the factory system brutally shattered these warm, human, and inefficient concepts and substituted for them the cold, sharp notion of 'competition.'"[30]

Part of the critique also came from idealistic but impractical advocates of a "back-to-nature" movement. Drawing extensively from the Romantic views of Goethe and Wordsworth, as well as vaguely-understood oriental religious views, the American Transcendentalist movement questioned the increased gulf between man and nature, which was an explicit goal for industrialization.[31] Emerson's landmark 1836 essay, *Nature,* promoted a pantheistic harmony between man and nature through a common spiritual experience in a unifying Oversoul. A highly mechanized society, he complained, wrenched man out of his natural origins. Humanity could recover its fullness only through a mystical immersion in nature. Later this creed would give rise to Victorian sentimentality, with its worship of the picturesque natural world. Henry David Thoreau's *Walden,* based on a year and a half spent exploring self-sufficiency in a quasi-domesticated wilderness, is the other major Transcendentalist testament. Like *Nature, Walden* is one of the early American statements of an environmental philosophy. It argues for the purity, simplicity, and clarity of human life and in favor of escape from modern civilization and a return to nature.

But the Transcendentalists failed dismally with their utopian Brook Farm experience in New England, partly because their philosophy was so uncompromisingly antimaterialistic, stressing the spiritual aspects of a return to nature and often a quasi-mystical union with God or a universal spirit. Tocqueville echoed this Transcendentalist critique when he wrote of the dangers of an "ennervating materialism" that could destroy the American character if it became preoccupied with industrialization. A bridge can easily be seen between these early antitechnology, nature advocates and the modern ecology-oriented environmentalists through the writings and work of George Perkins Marsh and John Muir.[32]

But even the most embittered critics of industrialization could not deny that the new steam-powered factories created a dazzling prospect. Humanity seemed embarked on an historic and revolutionary move from perpetual scarcity to an economy of abundance. John F. Kennedy spoke of a "rising tide which lifts all boats." Despite job loss or displacement, despite the filth and ugliness of industrial towns, and the hardships of factory life, workers sought out factory employment. They rushed from the life of the farm to the new factory cities. Steam power, the new machines, and the new organization of work promised abundance and created a new atmosphere of hope.

Yet the great riddle of the modern age of energy persisted. Industry was expected not only to produce profits and jobs, but to conquer the three archenemies of civilization: poverty, ignorance, and disease. Nevertheless, large numbers of people still lived under conditions of poverty, hardship, insecurity, and hopelessness. The critique of Western industrial capitalism was led by socialists, some Marxists, labor reformers, and utopians. This criticism received wider attention in the twentieth century and became part of the West's preoccupation with its own decline.[33] A new "holistic" or ecological approach claimed that a profit-oriented system could not achieve the intricate harmony between energy, technology, and society that was now required. Critics like Lewis Mumford, Jacques Ellul, and Ian McHarg denounced the "cowboy ethic" and the "Faustian bargain."[34] American industry was said to have forced a consumer "cargo-cult" psychology upon the mass of people, who had come to worship the false gods of growth and material gain.

THE CONSERVATION MOVEMENT BEFORE 1973

Some historical precedents exist for modern energy conservation, although America is hardly the birthplace of energy frugality. The first signs of a conservation ethic appeared early in the nineteenth century due to problems with soil exhaustion and replenishment in the East. Later, more widespread interest appeared as water resource management became critical in the far West.

Modern environmentalists can claim Romantic and Transcendentalist forebears ranging from Emerson and Thoreau to the later activities of Catlin and Muir. Muir's formation of the Sierra Club in 1892 marked a turning point in environmental activism,[35] because it represented a broad challenge to the assumption that industrialization was bringing "progress." Muir and his followers argued that wilderness was at least as necessary to life as urban civilization and that industrialization was corrupting mankind by substituting artificial wants for natural needs.

The movement was divided, however, by a conflict between "preservationists" and "conservationists." Preservationists wanted nature to be left alone; underground fossil fuel resources in wilderness regions were to remain untouched. This wing of the movement urged the development of the national park system, which made it a responsibility of the federal government to keep selected places off limits to energy development.

The conservationists first came into their own during the Progressive era, mainly due to the efforts of Gifford Pinchot and Theodore Roosevelt.[36] Prevailing attitudes in the departments of Agriculture and Interior, stressing multiple land use based on economic need and national strategic priorities, were shaped by the conservationists. Soil conservation is still a major activity of the USDA. Forest Service land management, combining use for public recreation and for private enterprise in the lumber industry, is a major conservationist activity. Beginning with the reform efforts of President Cleveland, who in the 1880s boasted that he had recovered 80,000,000 forest acres, the conservationist approach dominated American attitudes towards natural resources.

Responding to the conservationists, politicians launched a federal program to withdraw lands and resources critical to national security, beginning with presidents T. Roosevelt, Taft, and Wilson. The resources included phosphates and coal and Naval Petroleum Reserves in California, intended to fuel warships in time of crisis. Fears were also expressed that the nation's oil supply would be depleted in two generations because of wasteful extraction methods.[37]

Precedents less favorable to conservation were set during the Taft administration in the use of water power sites for hydroelectric development. Federal control was based on the principle of regulation-by-leasing, but the power industry sought long-term unregulated leases with virtually automatic renewal. As the public clamored for more electrical power, the 1901 Right-of-Way Act did allow the Department of the Interior to grant permits for power transmission lines across public lands and to build dams on public land sites, with the right of revocation. A water power bill in 1920 provided for long-term (50 year) leases with regular payment to the government.[38] On a state level a multiple-use trend was established at the turn of the century in California when the pristine Hetch-Hetchy Valley, part of Yosemite National Park, was nevertheless turned into a major water reservoir for San Francisco after a lengthy national controversy.[39]

Significant energy conservation did not come to the United States until the 1970s, when it was induced by oil shortages and rising prices. Regional natural gas shortages in the winter of 1977–78 also led to reallocations of limited supplies to residential areas and compelled industrial conservation and shorter work hours. In these emergencies, energy conservation was defined in ad hoc terms as a means to extend oil and gas supplies to forestall anticipated shortages and to minimize hardships. More recently, there are signs that a more long-term commitment to conservation may be taking hold, perhaps marking a historic transition from prodigality to frugality in the use of energy resources. These signs include the adoption of mandatory speed limits and improved average automobile mileage in new models and dramatic voluntary cutbacks in consumption by consumers reacting largely to higher prices.

THE LAST TWENTY YEARS: NEW FORCES

Historically, American energy development took place in a climate of relative harmony between the energy industry and consumer interests. Industry was anxious to promote the use of energy, and consumers were equally eager to use more and more energy, as long as prices remained low. Prices were kept low partly by efforts of industry to develop and exploit domestic and foreign resources and partly by political interference with market forces—in the form of subsidies for power projects and regulation of interstate natural gas prices. From time to time, strikes, especially in the coal industry, produced temporary shortages, but there was rarely a general concern, in industry, government, or public consciousness, with the long-run consequences of escalating energy demand. In 1952, the Paley Commission issued a prophetic warning of impending difficulties, including the likelihood of overdependence on imported oil, but it fell on deaf ears. Industrial priorities centered on exploitation of resources but not on conservation, efficiency, and the search for alternatives to nonrenewable resources. Profitability, productivity, stability, free-market mechanisms, and a sympathetic government and public were more important goals. Over the last 20 years this attitude of business-as-usual has been jarred by new and unexpected developments.

Environmental Awareness

Among the new forces, the first to make itself felt was a strong environmental awareness. The opening shot of the modern environmental movement was not directed against energy consumption. It was Rachel Carson's 1960 critique of the use of DDT, which launched the first major public debate in American history about an inherent defect in an important and useful technology.[40] In this controversy, the public first discovered technological sin. Carson's emphasis on the harmful effects of DDT not only put the question of technological effectiveness in a larger social context but suggested that the thoughtless application of a host of modern technologies, including new drugs as well as new chemicals, might be producing more harm than good. There were overtones of guilt and evil such as those that now revolve around the use of nuclear energy. The use of DDT for pest control had been considered a necessity by farmers, but the new environmental awareness, together with intense pressure from organized public interest activists, led to judicial, administrative, and legislative action resulting eventually in the banning of the use of DDT.

The environmental movement quickly broadened its critique to include a wide variety of the side effects of industrialization. Air pollution first gained major attention when a deadly killing smog hit Donora, Pennsylvania in 1948. When pollution from burning coal was identified as the culprit, environmentalists had ammunition to gain passage of the Clean Air Act of 1963 and the Air Quality Act of 1967. The National Environmental Policy Act of 1969, with the later creation of the Environmental Protection Agency (EPA),[41] was the single most important achievement. EPA established a policy that favored health and welfare-based standards as opposed to traditional standards based on technological and economic criteria. "Technology-forcing" laws included the Clean Air Act of 1970 and the 1972 Federal Water Pollution Control Act, both with explicit goals, timetables, procedures, and methods. For the EPA, as the standard-bearer of the environmental movement, there were six problem areas: air, water, pesticides, radiation, noise, and solid wastes. Americans were to have clean air by 1975 and clean water by 1985. Other notable actions included the Wilderness Act of 1964 and the numerous automobile pollution laws of the 1960s and 1970s. More recently, due to the pressures of economic decline and the energy situation, as well as a changing political climate, major environmental controls have been eased. Pollution control costs per person in the United States rose from $47 in 1974 to $187 in 1977, with job losses estimated at 20,000 and adding as much as a half percent on inflation annually. In reaction, the Clean Air Act was substantially amended by Congress in 1977 in order to reduce pollution control costs for industry.

Before 1960 environmental concerns over pollution, conservation, waste, despoliation, and scenic preservation generated mostly local or special interests. But by April 22, 1970—Earth Day—a powerful and sophisticated environmental awareness had become embedded in the national conscience and had made a profound influence on public policy. In the past, the quadrennial presidential campaign platforms rarely had an environmental plank. In recent times, a statement on the environment has become standard. Environmental impact statements and interventions were unknown a decade ago. Today, environmental protection costs are a significant factor in the balance sheet of virtually every major industry.

An Aging Industrial Plant

Some tensions and polarities that prevent resolution of the energy problem result from a certain lack of resiliency and flexibility in America's economic capacity to respond to challenges. The United States, its industrial plant undevastated by World War II, now has what is probably the oldest industrial structure in the world. In view of the rapid pace of technological change, the United States can be described in economic terms as a prematurely old nation. The patterns of capital development and technological implementation did not have to be fundamentally changed to overcome the devastation of war. Long-term undisturbed technological and industrial development was once a desirable and beneficial condition, but in time it can lead to stagnation. There are probably many reasons for the recent decline in American productivity, but one of them certainly is that the nation is burdened with outmoded plants and equipment.

When there is a widespread recognition of crisis, as occurred in the defeated countries after World War II, major social groups have an incentive that is otherwise lacking to cooperate for the common good. In such an environment, the industrial system can develop resiliency in high risk situations by dispersing or minimizing the effects of the crisis. In contrast, an industrial system beset by conflicting social demands, creaky in its plant, and unable and unwilling to innovate cannot easily respond to economic challenge, especially when the challenge has only a gradual impact. Hazards are intensified and localized rather than dispersed. There is less ability to deal with surprises and the unexpected, and short-term answers end up having long-term negative results. In the case of energy, the result is that transition to more secure and efficient patterns of energy use is likely to take place only slowly and without the universal and concerted effort necessary to maintain industrial leadership in an increasingly competitive world.

Risk and Safety

Another new feature of contemporary life is the widespread and intense preoccupation with personal safety.[42] This interest in safety is in part the result of the country's very success in developing and introducing new technologies. In the past Americans would tolerate occasional derailed steam engines, exploding steamboats, and the intermittent and very gradual improvement of automobiles, airplanes, and appliances. Today the situation has changed considerably, in terms of both technology and social attitudes. The pace of technological innovation is more rapid than in the past. Different standards for social acceptability have emerged. There may be no such thing as a perfectly risk-free technology, but minimally-acceptable safety standards are apt to be higher today than they were in America's past.

Despite all the defensive rumblings about the limits of technology, the American public still expects American industry to guarantee product safety and reliability. This consumer demand is based on a historic trust in technological "fixes" as solutions to technological problems. Americans tend to believe that industry is quite capable, if compelled, to produce safer products at no great increase in cost. Such expectations have been raised because technologies have historically become progressively safer and cheaper. Unfortunately, the reality is more complicated. Some improvements in safety can only be introduced at significantly higher costs—costs that consumers may be unwilling to bear and that handicap American producers in foreign markets. Difficult tradeoffs must therefore be made in order to satisfy both the new demand for safety and the older expectation of continuous technological improvement.

Stagflation

A third new feature is the phenomenon awkwardly called "stagflation." Certain combinations of economic conditions, such as inflation and recession, were historically supposed to be mutually exclusive, or mutually corrective. Instead they have now combined to create a seemingly intractable economic problem. As a result, the conventional view of the link between energy development and economic trends is open to serious question. One conventional doctrine of industrial efficiency had it that intensive energy use lowers other costs of production. Now that energy costs are rising, this no longer holds. Formerly, increasing prices for energy services, coupled with the increase in demand that could be taken for granted, would stimulate investment in new capacity, but higher capital costs and uncertainty over future demand now inhibit such investment. Productivity, moreover, no longer seems to be so much a function of higher energy use; at some point it seems to require restructuring of organization and incentives rather than simply a more efficient use of energy and other resources.

The complexities of stagflation are awesome. While Americans are just coming to grips with it, European and developing nations have long known what it is to live beyond one's means. Savers and investors have lost faith in the future—the very incentive for saving and investing. Industry avoids long-term capital investment in new plants, or in research and development, the condition of new opportunities. Yet effective answers to the energy crisis cannot be found in the growing dependence on short-term low-risk projects with quick returns. Stagflation engenders managerial attitudes incompatible with the long-term responses required to meet the energy problem.

Historically, both economic stagnation and inflation have proved to be only temporary problems, but the new economic climate may well be more deep-seated. For the energy problem, the persistence of stagflation could be especially serious. Energy-intense technologies historically have depended upon growth and progress for their future development. The effect of stagflation upon new investment in energy resources threatens to upset this interdependence.

Government Regulation and Intervention

A fourth new feature of the modern American social landscape is widespread government regulation and intervention in the economy. The growing role of central government has been a major aspect of American economic life beginning with the New Deal. It reached truly epic proportions in the 1960s and 1970s. For many economic actors it became a major axiom that success or failure begins and ends in Washington, due to the role of federal procurement, subsidy, and regulation. Nowhere is this more evident than in federal policy for energy and natural resources. Producers and consumers, organized interest groups and the military, ethnic minorities, the rich as well as the poor, all expect government to intervene at the first hint of trouble for any of them. The energy crisis has not hurt as much as it might have because government agencies have gone to great lengths to cushion its impact, on the nation as a whole and on particular regions and social groups. The result has sometimes been insulation from harsh reality. Buffered against otherwise "uncontrollable" forces, all groups tend to rush for protection to government, however ready they are to denounce bureaucratic bungling and legislative incompetence. Government energy policy "fixes" are viewed as substitutes for technological "fixes." It is said that the problem is not scarcity but management. That is, policy decisions, court cases, legislation, and energy agency actions are the means to stimulate production of more oil and natural gas and solve the nuclear and solar debates and to guarantee the American way of life. Unfortunately, this attitude only makes the problems worse because it shifts the burden of responsibility onto government agencies and politicians who are powerless to control market forces and foreign developments and who, in the last analysis, can only act decisively if there is a strong public consensus for a particular course of action. In the absence of such a consensus, reliance on paternalistic government is bound to lead to disappointment.

3 The Debate Over Energy Policy

While energy supplies were still plentiful and affordable, most Americans had no incentive to concern themselves with energy policy. As a result of recent shortages of oil and increased prices for all forms of energy, indifference has given way to apprehension, which has triggered a wide-ranging and sometimes highly charged debate over energy policy.

The issues raised in this debate are by no means all new. In 1918, for example, the brilliant mathematician and chief engineer of the General Electric Company, Charles Proteus Steinmetz, read a paper[1] to a professional society, the American Institute of Electrical Engineers, in which he surveyed the country's potential supply of hydroelectric power and fossil fuels. He concluded that for the foreseeable future, the best use of these primary sources of energy would be to produce electricity—not a surprising position for someone whose company had a keen interest in the marketing of electrical appliances. But Steinmetz was too good an engineer not to recognize that current practice was often profligate, especially insofar as "many millions of kilowatts of potential power are wasted by burning fuel and thereby degrading its energy." Most of these lost kilowatts could be recovered, he pointed out, "by interposing simple steam turbine induction generators between the boiler and the steam heating systems, and collecting their power electrically"—the technique now known as cogeneration and common in European countries but still rare in the United States.

In the same article, Steinmetz also discussed another source of energy that he believed would one day become essential: "A source of energy which is practically unlimited, if it could only be used, is solar energy." He calculated that the solar radiation falling on the United States was the equivalent of 800,000 million kilowatts, a thousand times as much as was contained in the chemical energy of the coal that would be consumed in 1918, and 800 times the country's hydroelectric potential. Assuming that land not suited for agriculture could be set aside for collectors "and assuming that in some future time, and by inventions not yet made, half of the solar radiation could be collected, this would give an energy production of 130,000 million kilowatts." If even a tenth of solar radiation could be put to use, Steinmetz estimated, the energy supplied "would be many times larger than all the potential energy of coal and water. Here then would be the great source of energy for the future."

From time to time, other equally astute and prescient commentaries of the same sort appeared in the technical and popular literature. In 1956, to cite a more recent example, the petroleum geologist M. King Hubbert made what was then considered the overly pessimistic prediction that domestic oil production would peak between 1966 and 1971 and then decline sharply. (The peak was reached in November 1970.)[2] But like Steinmetz before him, Hubbert attracted little popular interest, and his warning had little if any political impact. In the federal executive and Congress, policy for energy was decided piecemeal under the heading of "fuels policy" rather than comprehensively, and with little regard for long-term trends. Each energy source was treated separately within the Department of the Interior and in Congress, enabling clusters of regional, industrial, bureaucratic, and consumer interests with special concerns for coal, oil, natural gas, and electric power to shape policy as major issues arose. For security reasons, responsibility for fostering the development of atomic energy was placed in a special new agency, but this decision to treat yet another energy source in isolation from all others conformed to the traditional pattern.

A new awareness of the need to define an energy policy did emerge for a time, but without much effect, in the years immediately following the end of the Second World War. It was keenly recognized by many at high levels of policy making that resource scarcity had played a considerable role in the Japanese decision to initiate war and in the breakdown of the German military machine in the last months of the war. Out of concern for the reliability of America's energy resources in the event of a future emergency, President Truman ordered a major study to be made of the future availability of energy supplies by a commission chaired by broadcasting executive William S. Paley. In 1952, the commission produced its report, *Resources for Freedom,* which warned that at current and expected rates of depletion, domestic fossil fuels would rapidly become scarcer and more expensive. Noting that "it took nature over 500 million years to store in the ground these stockpiles of 'fossil fuels' which civilization is now consuming in a flash of geological time," the report proposed that "such unconventional sources as solar and atomic energy" be developed to replace fossil fuels in the near future. The report also recommended that the United States assist other, poorer countries to explore for new sources of fossil fuels, but warned against overdependency on Middle East oil. Greater emphasis was urged on encouraging technological improvements, in order to make it easier to substitute fuel sources and to recover more oil and gas from known deposits. A gradual shift in patterns of consumption was forecast back to coal from oil and gas. More attention was urged to the commercial development of synthetic liquid fuels from shale and eventually coal. Although the report did not press the government to take a major role in implementing these proposals, it did recommend the formulation of a "comprehensive energy policy" and a single focus for administrative review of energy issues in a single department.[3]

By the time the Paley Report was issued, however, the Korean War had diverted the attention of the policy makers to short-term needs, and before the report's proposals could be digested politically, a new Republican administration had come into office strongly averse to anything that smacked of government planning and equally strongly committed to allowing the price mechanism to determine rates of resource extraction and patterns of allocation. In practice, the Eisenhower administration was sometimes much more activistic than its ideological professions allowed for. The president and his Secretary of State, John Foster Dulles, intervened to assure continued access to Iranian oil for American companies and to resist Iran's effort to nationalize foreign holdings. Eisenhower's "atoms for peace" plan stimulated domestic and foreign business for the fledgling atomic energy industry. Contrary to the recommendations of the Paley Commission, however, no steps were taken to formulate a comprehensive policy or to reorganize the administration of energy issues to provide greater coordination of policy.

With the Paley Report's recommendations largely ignored, the emphasis of private and public efforts for the next two decades was concentrated on expanding supply by drilling for oil and gas and increasing imports of oil and on building new generating capacity to meet existing and expected growth in demand for electricity. Coal production declined as coal became less competitive with oil and natural gas. Price controls kept the price of natural gas low and oil use was stimulated by a variety of incentives to producers, including the oil depletion allowance and a provision enabling oil importing companies to offset foreign royalty payments against domestic tax liabilities. The policies adopted were often contradictory in their purposes and effects. While some encouraged importation, others sought to limit imports in order to protect small domestic "independents." One agency sought to foster atomic energy while another set pollution standards that delayed construction and imposed severe economic hardships on the industry. The net result, however, was that far from experiencing the shortages predicted by the Paley Report for the near future, the country actually experienced a glut of energy supplies, accompanied by declining prices.

With the benefit of hindsight, it is clear that this era of plenty was the result of anomalous and temporary conditions. Especially in the earlier years of the postwar period, while other industrial countries were still struggling to recover, American consumers faced only weak foreign competition in oil markets. Policy decisions that effectively raised consumption and depleted supplies domestically, while increasing dependence on foreign sources, only served to postpone the advent, and increase the severity, of exactly the problems to which the Paley Report tried to draw attention. As consumers grew more and more dependent on declining domestic reserves and increased imports, the nation became more and more vulnerable to shortages and cutoffs of supply.

Misgivings about the long-term effects of growing reliance on diminishing fossil fuel reserves were often allayed by rosy predictions of the impending benefits of atomic energy. In due course, it was thought, atomic energy would come to the rescue of consumers used to relying on fossil fuels. Gradually, more and more nuclear plants would replace conventional generating facilities and provide even cheaper energy that would soon prove to be virtually inexhaustible, especially once breeder reactors and, eventually, fusion systems became commercially feasible. Neither corporate-sector managers, driven by concern for short-run profitability, nor government overseers, concerned with maintaining orderly and competitive markets and protecting consumers against monopolistic practices, saw a need to restrain the appetite for increased consumption. In the executive, this complacency reached its nadir in what has been called the policy of "benign neglect" followed in the early years of the Nixon administration before the 1973-74 oil embargo.[4]

As this complacency was rudely shattered by the oil shortage resulting from the embargo and the steep increases in prices resulting from subsequent decisions by the oil-exporting companies organized in OPEC, belated efforts were finally made to develop national policy and the questions at issue began to be widely debated.

THE NATURE OF THE DEBATE

The energy debate of recent years has been many-sided and complex, as a brief overview of the actors, goals, policies, and implications will help to indicate.

Actors

Among the principle participants in the debate are such interested parties as the oil companies (both the large multinational companies and the smaller independents); investor-owned, publicly regulated electric utilities; coal mining companies (some of which are now part of larger energy conglomerates); states and regions with actual or potential fuel resources; companies that produce nuclear and other power-generating facilities; the various federal agencies with supervisory and regulatory functions pertaining to energy; experts in various technologies, in the economics of energy supply and demand, and in foreign policy; organized groups of citizens in favor of certain options and opposed to others; and of course also countless concerned citizens and their official and unofficial spokesmen in politics and the media.

Goals

To some of the participants, perhaps most, the goal of a national energy policy must be to seek to assure continued supply of energy services, whether by conventional or unconventional means. To others, however, the goal is to adjust patterns of consumption, and with them patterns of expectation, in order to manage the transition from reliance on fossil fuels to reliance on renewable resources. There are grounds of agreement between these two views, but more often it is the differences that tend to receive the most emphasis in the debate, as is nowhere more evident than in the argument over the rate of economic growth that is to guide energy planning. Those who hold the first view tend to favor relatively high rates of growth; those who hold the second tend to favor relatively lower rates of growth, or if not to favor them, then to argue that they are all but dictated by energy constraints.

Policies

The approach to policy favored by the participants in the debate is often shaped by attitudes toward government intervention in the economy. Those who favor a significant degree of intervention sometimes advocate the nationalization of major energy companies or at least efforts to increase regulatory control of their behavior, by strict application of antitrust laws and safety and pollution regulations. They are also apt to favor active measures by government to cushion hardships suffered by the least advantaged sectors of society because of efforts to curb energy use by allowing prices to rise. On the other side of this divide are those who argue that government intervention in the energy area has only made matters worse, and that the best way to handle the problem now is to remove price controls and subsidies and lighten the burden of regulation consistent with necessary standards of public safety. In their view, the best course is to make it possible for the private sector to adjust to new demands both for new supplies of energy and for more efficient use of the supplies available. This issue is also raised in the controversy over nuclear power, but attitudes toward the relative safety and need for nuclear power are also in play.

Implications

Domestically, the energy debate touches on many aspects of American social structure, including standards of living, the democratic process, degrees of confidence and civility, and other similar issues, both tangible and intangible. In terms of foreign policy, the implications range from concern for the impact of energy policy on America's role in the world, on relations with allies, adversaries, and the developing countries, and on the dangers associated both with the proliferation of nuclear weapons and the tensions that might arise as a result of energy scarcity.

If there is a single set of questions about which the debate may be said to focus, it is probably the dispute over whether to encourage a broad range of efforts to curb consumption and at the same time promote all possible means of supply, or whether to institute more fundamental changes in consumption and technology in order to move as rapidly as possible away from reliance on fossil fuels and nuclear energy toward the use of small-scale systems relying largely on solar energy.

To understand this central debate, it is useful to consider its ideological implications, which have brought to mind an older American debate between Jeffersonians and Hamiltonians.[5] While this parallel is far from perfect, it is a useful way of thinking about the current debate. The new Jeffersonians advocate an energy policy designed to inhibit the growth of corporate power and big government, and instead to promote decentralization and participatory democracy. They urge the adoption of technologies that minimize dependence on fossil fuels and rely instead on renewable sources, or solar energy. They tend to oppose nuclear power on a variety of grounds, including its potential for promoting proliferation of nuclear weapons. They tend to sympathize with the demand of the developing countries for self-determination and for a global redistribution of wealth and political control of market forces, as envisioned in the call for a New International Economic Order. They stress the need for the advanced countries to assist the less developed countries in learning how to avoid high-technology, fossil-based energy solutions and to pursue ecologically sound, village-based schemes of energy development emphasizing the use of renewable and other small-scale sources.

The new Hamiltonians generally view most or all of these proposals as unrealistic because such proposals reflect a mistaken view that large-scale conventional technologies can easily be dispensed with in favor of new but untried alternatives that have yet to be shown to be feasible in actual operation. They argue that the advocacy of exclusive reliance on solar energy reflects a bias against continued economic growth and against the system of individual opportunity and free choice that such growth makes possible. In its place, they contend, those who urge a policy of low-growth or no growth would impose penury in place of affluence, mass conformity in place of individual choice and diversity. They also argue that a failure to maintain productivity by developing all possible sources of energy will weaken this nation's power to protect itself and its allies and will prevent it from providing the levels of assistance the developing countries will need to cope with the added burdens imposed by shortages of energy.

JEFFERSONIANS AND HAMILTONIANS: AMORY LOVINS AND HIS CRITICS

One of the most prominent and articulate participants in the public debate over energy policy has been Amory Lovins, a consulting physicist associated with the environmental organization, Friends of the Earth. His article, "Energy Strategy: The Road Not Taken?", in *Foreign Affairs* magazine,[6] attracted both enthusiastic support and impassioned criticism. In subsequent writings, Lovins has openly declared that his bias is "in essence, Jeffersonian,"[7] but he has also contended that while "nobody can make a completely value-free analysis."[8] the validity of his arguments does not depend upon acceptance of his value premises. This is certainly true insofar as his analysis concerns calculations of the costs, vulnerability, and other technical characteristics of different options. Nevertheless, his diagnosis of the energy problem aims to be comprehensive, and to this end he deliberately invokes social and political criteria of acceptability along with technical criteria. The "soft path" he advocates is not justified solely on the ground that small-scale systems using renewable sources of energy are less expensive, more efficient, and less vulnerable than the "hard path" alternatives, but also on the ground that they are more socially desirable. The "hard path" is opposed because it requires more centralization and is therefore Hamiltonian rather than Jeffersonian, in the sense in which Lovins understands the difference, as when he writes that "although humanity and human institutions are not perfectable, legitimacy and the nearest we can get to wisdom both flow, as Jefferson believed, from the people, whereas pragmatic Hamiltonian concepts of central government by a cynical elite are unworthy of the people, increase the likelihood of major errors, and are ultimately tyrannical."[9]

The link between this political belief and energy policy is established in a general way by the advocacy of decentralized systems, which diminish the economic and political power of large corporations, government bureaucracies, and technocratic elites in favor of small-scale enterprise and individual and community autonomy. In particular, it supports such statements as this comment on the human unacceptability of centrally generated electricity: "In an electrical world, your lifeline comes not from an understandable neighborhood technology run by people you know who are at your own social level, but rather from an alien, remote, and perhaps humiliatingly uncontrollable technology run by a faraway, bureaucratized, technical elite who have probably never heard of you."[10]

In arguing against the "hard path," Lovins contends against the view that he takes to be the conventional wisdom, or the attitude that has held sway until now and that continues to be supported by industrial interests and their supporters within the technical community. This path, he argues, involves an effort to expand supply indefinitely by finding more and more fossil fuels, and supplementing these fuels with nuclear power, in ever more costly, complex, centralized, and gigantic systems. He calls this policy one of "strength through exhaustion." It requires that "in essence, more and more remote and fragile places are to be ransacked, at ever greater risk and cost, for increasingly elusive fuels, which are then to be converted into premium fuels—fluids and electricity—in ever more costly, complex, gigantic and centralized plants."[11] The use of energy would double, even treble, over the next several decades. In the process, privately owned energy companies would prosper and grow stronger politically. Regulatory requirements would also reenforce tendencies toward greater bureaucratization in government and toward a more powerful role for technocrats. Public subsidies for large-scale energy projects, which Lovins estimates have reached an annual size of ten billion dollars, would continue and probably grow even larger.

The most obvious problem with an effort to continue along this "hard path," according to Lovins, is that it is simply "unworkable" economically and politically. Huge, unsupportable investments of capital would be required. Even if such investments could be made, moreover, they would aggravate unemployment because they produce fewer jobs than any other comparable investment. Politically, such a course is bound to meet with resistance from those directly affected by the pollution and safety risks large-scale systems inevitably entail. Opposition would also be aroused because of the "structural" implications of hard path solutions, including centrism, vulnerability, technocracy, alienation, repression, and other stresses and conflicts, including tensions between regions and income groups. Centralized systems are vulnerable because power lines can be severed by a single rifleman or disconnected by a few strikers, while nuclear facilities invite sabotage and the diversion of dangerous materials. They would therefore require "stringent social controls" as well as repression of "dissent against the energy system." As more and more people became unwilling to accept decisions made by established authorities, political alienation would grow and threaten the legitimacy of government.[12]

To this list of dire domestic consequences, Lovins adds an even more threatening chain of potential international consequences:

> We use the apparently cheap energy wastefully, and continue to increase our dependence on imported oil, to the detriment of the Third World, Europe, Japan, and our own independence. We earn foreign exchange to pay for the oil by runing down domestic stocks of commodities, which is inflationary; by exporting weapons, which is inflationary, destabilizing, and immoral; and by exporting wheat and soy beans, which inverts midwestern real estate markets, makes us mine groundwater unsustainably in Kansas, and raises our own food prices. Exported American wheat diverts Soviet investment from agriculture into defense, making us increase our own inflationary defense budget, which we have to raise anyhow to defend the sea lanes to bring in the oil and to defend the Israelis from the arms we sold to the Arabs.[13]

To the argument that only by maintaining high levels of energy supply will the advanced countries be able to aid the developing countries, he argues both that it is possible to decouple growth in output from growth in energy consumption and that "high growth in overdeveloped countries is inimical to development in poor countries."[14] He does not explain how this is so, but he seems to be suggesting that the "trickle down" theory works as poorly internationally as its critics say it does domestically. Presumably, the growing economic strength of the advanced countries makes them even better able to dictate the terms of trade to the developing countries and to prevent them from adopting measures to promote self-sufficiency.

Fortunately, there is an alternative—the so-called soft path. By this term Lovins means those energy technologies that are not complex and overcentralized but that match scale and end use to purpose. Instead of using hydroelectric dams to provide electricity for heating water—"cutting butter with a chain saw"[15]—this power would be used for industrial processes like steel making that require large amounts of power. Homes and hot water, however, would be heated exclusively by solar technology. Instead of trying in vain to build new generating capacity, at ever increasing marginal cost, the soft path would rely on conservation to reduce the need for energy by improving the efficiency with which it is used. The soft path would eventually lead to a lower ratio between energy consumption and production, without lowering expectations. Existing centralized power stations would be used as a bridge to a sustainable future, but they would not be replaced as they run down. In theory, at least, by the use of soft technologies "an affluent industrial economy could advantageously operate with no central power stations at all!"[16]

In a soft path strategy, development of many types of solar technology would be encouraged, including the use of agricultural wastes as biomass for conversion to liquid fuel but not large-scale systems such as those that require huge arrays of collectors in the desert or giant satellite receivers. These are classified as "hard" technologies. Similarly, the effort to generate electricity by fusion is deplored. Fusion would be more complex and difficult than even fast breeders, and it might produce fast neutrons that would be used for bomb material. Even if it turns out to be a clean way of developing energy, "we shall so overuse (it) that the resulting heat release will alter global climate. We should prefer energy sources that give us enough for our own needs while denying us the excesses of concentrated energy with which we might do mischief to the earth or to each other."[17] (Curiously, however, this self-denying ordinance is applied only to the production of electricity. Since there is no objection to the development of biomass liquid fuels, there would be nothing to prevent overproduction of these fuels, even though they too may lead us to do mischief.)

Nuclear power is objectionable not only because it is dangerous and promotes overcentralization but also because it makes proliferation of nuclear weapons much more likely. Lovins argues that a U.S. decision to forego nuclear power and to stop its own sales of nuclear power facilities to other countries would have an inhibiting effect on other suppliers and thus dramatically reduce the likelihood of proliferation.[18]

In order to reverse the present course, barriers in the way of conservation and more rational supply technologies would need to be removed. This would require, above all, revision of building codes that inhibit construction of more energy efficient buildings. Subsidies for conventional fuel and power systems would be reduced and eventually eliminated, and antitrust laws would be vigorously enforced, to prevent giant corporations using profits from rising energy prices to acquire control over a variety of energy sources. Market forces would be used to bring energy prices to the level of marginal replacement costs. To speed the process, an energy tax—in effect a "severance royalty"—might be imposed on all fuels, according to their energy content, as they come out of the ground or into the country.[19]

Measures aimed at conservation would include retrofits on existing buildings and new designs for energy efficient buildings, as well as an increase in the use of cogeneration. Soft technology aims to rely on renewable forms of energy—on "income rather than capital," on a diverse mix of sources, on flexible and relatively low technology systems, matched in scale and geographic distribution to end use needs, and in energy quality. For this purpose, the use of electricity for low-grade purposes is to be avoided, in favor of an "elegant frugality," marked by such values as thrift, simplicity, diversity, neighborliness, humility, craftsmanship...."[20] According to Lovins, soft technologies are "inherently more participatory" and less coercive than hard technologies: no one can opt out of nuclear risk, and in an electrified society, everyone's lifestyle is shaped by the homogenizing infrastructure and economic incentives of the system. Anyone anxious to live in an uninsulated house or drive a gas guzzler would be free to do so, provided he pays the social costs they entail.[21]

The viewpoint expounded by Lovins has been criticized by many commentators on various grounds. Many of the objections refer to technical and economic estimates related to different technologies. Lovins has frequently been accused of overestimating the costs and underestimating the benefits of conventional power while grossly minimizing the problems of introducing unproven alternate technologies. One industry critic argued that Lovins is not so much proposing a new course of action as trying to inhibit all action, by urging that we stop doing what we know how to do "in order to do something better that we don't yet know how to do."[22] Advocacy of soft technology, he has observed, is in effect a rationale for opposing everything with a realistic chance of alleviating the energy problem, such as Alaskan oil, western coal, hydroelectric dams, offshore oil, geothermal energy, nuclear energy, and shale conversion and other synthetics.

Another industry critic, Sheldon Butt, president of the Solar Energy Industries Association, has denied Lovins's claim that all the heat required for most buildings can almost always be supplied by solar systems at a cost that compares favorably with the marginal cost of power generated by other means. "This," according to Butt, "is simply not true." Solar energy "cannot reasonably be a 100% source of heating energy." Lovins has cited Danish data to show that with systems designed to store heat in water, 100 percent solar heating is economical. Butt has contended that the capital costs of storage facilities would be higher than storing an amount of coal equivalent in energy in a central power station. He predicts that no more than 12 percent of the energy budget could possibly be solar by 2025,[23] in contrast to Lovins's contention that with an adequate effort to develop the technology and a removal of subsidies to other sources, solar would account for much more by that time.

Two energy economists have charged that Lovins's economic assumptions ignore the likely effect on demand of relatively stable prices. Theoretically, at least, efforts to hold energy consumption constant while providing relatively inexpensive energy are likely to be frustrated by increased demand.[24]

Other critics have charged that Lovins is trying to use energy policy as a lever of social change. Although he responds by claiming that he is simply following the logic to which his analysis leads,[25] they argue that the analysis proceeds from a predetermined conclusion. A social theorist has suggested that Lovins seems to think that by making the right technological choices, people can solve certain long-standing social and political problems, free themselves from the domination of corporate interests, bureaucracy, and experts, and in the process restore power to local communities and neighborhoods and render complex problems more intelligible and more manageable at the local level. Lovins has replied that while he does not suppose that the soft path will solve such problems, "at least it should prevent such problems, in the context of energy policy, from getting worse."[26] The critic contends that Lovins's prescriptions fall in the same category as other earlier efforts to promote settlement houses and community self-help movements—efforts that cannot solve large problems and tend to be replaced or absorbed by more efficient, more professionalized, and large-scale social systems.

While many of the critics agree with Lovins's stress on the desirability of energy conservation, they question whether conservation alone can allow for continued economic growth at a high enough rate to satisfy aspirations and enable the nation to maintain its power in the world. They suspect that Lovins views conservation as a way to force a change in habits of consumption and to impose limits on American power—on the ground that energy surpluses tempt people to do harm to the earth and to themselves. Although Lovins argues that he favors using market forces to encourage conservation and to stimulate the development of soft path technologies, his critique of big business and oligopoly is interpreted by some critics as a veiled attack on industrial capitalism. One of them sees in Lovins's preferred new society nothing but "a society of peasants and craftsmen."[27] Although he ridicules such charges, Lovins lends credence to them in such cryptic remarks as his claim that under modern conditions the Jeffersonian and the Maoist philosophies merge in pointing to the underlying economic problem of which the energy problem is presumably an outgrowth:

> As we learn to question the ability of present policy to serve both public and private ends, the legitimacy of those ends themselves comes up for review. Our know-how has far outstripped our know-why; and as we seek to redress the balance, old political concepts begin to reassert themselves. Grassroots democracy acquires a more concrete meaning. Jefferson and Mao gain a curious affinity.[28]

Is this merely traditional American populism joined to a once fashionable view of Maoism as nothing but Chinese "agrarian reform"? Or does it suggest something more radical, such as the idea that social reform must be aimed at undermining the structure of corporate capitalism in favor of experiments with backyard smelters and communal enterprise, and perhaps also a redistribution of wealth? Does the participatory democracy Lovins calls for include a tacit encouragement of acts of civil disobedience against nuclear power facilities, and perhaps against all other "hard path" technologies? Is that why Lovins—who is normally careful about his choice of words—fears that "dissent" against the energy system may be repressed?

His critics sense that Lovins speaks not only for alternate technologies but for an alternate set of values. Butt voices a common concern among Lovins's critics when he suggests that "the basic justification for our technological society, as it has evolved over a period of somewhat less than two centuries, lies in the exceptionally broad opportunity which it has provided to the citizens for individual expression and for the expression of political initiative."[29] While Lovins cannot fairly be accused of wanting to restrict citizens' political initiative, at least one critic wonders whether the policies he advocates would not require the restriction of economic initiative. "At no time in history," Butt observes, "have more individuals had the opportunity to devote a large proportion of their labor to those things which they individually elect as expressing their own desires, after first having satisfied their own needs." This opportunity, he adds, depends upon a system of "individual rewards."[30] It is interesting that in commenting on this criticism, Lovins avoids committing himself either for or against the belief in equal opportunity for unequal reward. Instead, he criticizes Butt's view as an instance of the chauvinism of people in industrially advanced societies, and observes that anthropological studies show that so-called primitive peoples "have considerably more leisure and more surplus resources to devote to their highly developed arts than do typical members of the most affluent Western societies today."[31]

In international terms, Lovins claims that he is merely urging an effort to minimize the dangers of nuclear proliferation and to improve the prospects for international equity, cooperation, and development in the poorer countries. Although he seems to imply no wish to promote fundamental changes in socioeconomic systems or in the balance of power, his strong advocacy of the New International Economic Order proposed by the bloc of developing nations suggests that he would be ready to countenance the replacement of market mechanisms and the role of multinational corporations by a more political, more bureaucratic system, and to require a redistribution of wealth among nations.

The argument between Lovins and his critics, then, is evidence that the general public debate over energy policy in this country can in some respects be thought of as a revival of the argument between Jeffersonians and Hamiltonians. Inevitably, however, historical changes make the simple repetition of old arguments in new circumstances awkward at best and worse when they lead to self-deception. Jefferson and Hamilton were arguing about whether the United States should move on an agrarian or an industrial path. Jefferson was not calling for the "repeal" of an industrialization that had already taken place. Would he have wanted to undo an industrial society created under the aegis of democracy and providing for the pursuit of happiness on a scale even he could scarcely have imagined? Would he have wished to limit individual choice so that people would not be free to harm themselves or others? We can be sure that he would have sympathized with efforts to preserve local autonomy and "grassroots" democracy. We can be much less certain whether he would have found large-scale industrial capitalism an enemy of individual autonomy, whether he would have been in favor of strong government to counterbalance the strength of large economic interests. We can only be sure that he would have wished to preserve the fundamental principles of private ownership and exchange, for Jefferson was no communist, neither Marxist nor Maoist.

As to Hamilton, is it really true that he was in favor of a "cynical elitism," as Lovins claims? Hamilton certainly feared mob rule and believed that society should be guided by its most able men—those in whom the spark of civic virtue could be nurtured into the capacity for statesmanship and public service. He was content to allow the best to achieve a privileged status because he believed, in company with others nurtured in eighteenth-century ideas, that the best way to assure the common good was to allow for the free play of talents and interests. Jefferson also believed in the need to nurture a "natural aristocracy" as the governing class of a free society—an aristocracy that was to be "raked from the rubbish"[32] by a system of public education. Both of them were thinkers of a democratic tendency—republicans, in the then preferred usage; they differed in the confidence they had in the capacity of the common man to regulate his own affairs and contribute constructively to affairs of state and in the role they were prepared to give to central public authority.

What either of them would make of a society as populous, complex, and highly organized as the United States has become, and with its role in world affairs, is anyone's guess. Would they have differed as Lovins as his critics do over energy policy? Perhaps, but it is hard to envision either of them, if apprised of the modern choices, turning his back on the complex interdependence of modern civilization in favor of a return to a simpler era when each subcommunity could aim to provide for its own needs from its own resources and to make itself as independent as possible of all others.

Whatever happens to America's energy systems—whether present emphasis on central generation of power persists or is replaced by smaller units relying on renewable sources—we can hardly suppose that technological changes will automatically bring profound alterations in social values. In some respects the choice of energy technologies does, as Lovins points out, either serve or disserve particular values. It does not follow, however, that the best test of the linkage is whether the technology is close or remote, small or large, and whether the resources used are exhaustible or renewable. As food industry spokesmen are fond of pointing out, the nutritional content of processed food is often as high or higher than that of organically grown foods. To the domestic consumer of energy it may not matter how the electrons get to his home, or whether they are produced by the burning of fossils or the heat from solar radiation. What matters most to him—and to the stability of the social system—is that the energy services he depends upon continue to be available. It is arguable that insofar as large-scale energy systems using a variety of resources, provide abundant energy, they help make possible a degree of individual autonomy and stable self-government that both Jefferson and Hamilton sought to assure. Only to the extent that these systems have offsetting disadvantages is there a basis for serious concern.

ENERGY AND IDEOLOGY: THE PUBLIC DEBATE OVER POLICY OPTIONS

The controversy sparked by Amory Lovins brought ideological considerations to the surface of the public debate over the energy crisis. For the most part, however, this debate has been preoccupied with arguments over policy in which ideologies are entangled with economic interests and technical judgments. Nevertheless, a review of the policy debates will show that ideological considerations play at least as important a role as these other factors in shaping and constraining policy decisions.

Although Americans are not as overtly committed to ideological doctrines as people often are in other political cultures, their views of the aims of social life and government have been profoundly influenced by the ideological principles invoked originally to justify the American Revolution and subsequently institutionalized in public law and the political system.[33] Appeals to these principles are sometimes only of symbolic significance, as in the case of Fourth of July oratory, when the intention is to reenforce patriotic sentiment. At other times, however, especially when the aim is to mobilize support for political parties, politicians, and policies, the appeal to ideological principles is much more meaningful, because these principles can shape electoral behavior and constrain the choice of policies.

Traditionally, the ideological principle most frequently identified as the cardinal element of the American political creed has been the belief in the liberty of the individual. Except during wartime, this belief has manifested itself in an aversion to centralized, concentrated, and comprehensive authority. The generally accepted view has been that in order to protect the liberty of the individual, the power of government must be held in check. During the Progressive era, and still more during the New Deal, this traditional belief was challenged by those who contended that in view of the growing concentration of economic power and the need to coordinate an increasingly complex society, it was necessary to assign to government a positive responsibility for economic management and for the promotion of social welfare. In a number of respects, their efforts to modify the traditional suspicion of strong government succeeded in bringing about important changes.

In the case of energy resources, one result was to strengthen the bargaining power of coal mine workers against owners. Another was to make provision for public works projects designed to provide affordable energy in rural areas and in regions where energy might promote economic development, notably in the area served by the Tennessee Valley Authority. In the aftermath of World War II, the reforms of the New Deal period were not extended further, and a kind of equilibrium seemed to have been established. Government agencies came to play a significant role in stimulating and regulating development, but the actual provision of energy services was left to private industry and regulated public utilities. Thus, the Atomic Energy Commission was established to supervise the production of atomic weapons and to encourage the application of nuclear energy for civil purposes. Federal agencies also acquired responsibilities for assuring mine safety, for regulating prices of oil and natural gas, and for setting import quotas for oil. Tax policy was adjusted to provide incentives for oil drillers in the form of depletion allowances. Foreign policy was also brought into play to protect access to foreign sources of oil on favorable terms.

This combination of government oversight and private enterprise had two important results. Energy prices were kept relatively low, with the result that economic growth was given a powerful stimulus, and the nation became more and more dependent on oil and natural gas. Because much of the oil was imported, the nation became vulnerable to precisely the kind of action taken by the Arab oil-producing states in 1973. The embargo, coupled with the strengthening of OPEC, succeeded in imposing a dramatically higher price for oil, with repercussions not only for the American economy but for national security as well. Among many Americans, however, the initial impulse was not to see the crisis as the result of imprudent public policies but instead to try to pin the blame on the oil companies, for creating a supposedly artificial shortage, or on the federal government, for interfering with market forces to keep energy prices artificially low, or on the OPEC countries, for colluding to control supply and price. (The leaders of the oil-producing countries, in their turn, blamed the profligacy of western consumers and sanctimoniously asserted that they were merely righting old wrongs and compelling the world to value its finite supplies of fossil fuels more realistically.) Considerable effort was expended in these recriminations, before public attention came to focus on practical proposals for dealing with the problem.

Although some of the proposals sought to address the problem by nationalizing or breaking the power of the giant energy companies, these made no headway, partly because it was not obvious that they would have any positive effect on the immediate problem, but also because they cut against the grain of conventional ideological conviction. More in keeping with ideological convention, though scarcely more realistic, was President Nixon's announcement of a plan to achieve "energy independence." The plan called for the adoption of policies aimed at making the nation potentially self-sufficient by the end of the decade. This was to be accomplished by allowing prices to rise—which would lead to decreased consumption and increased domestic production—and by expanding nuclear power capacity to replace reliance on oil and natural gas. The Ford administration sought to continue in the same direction, envisaging the construction in ten years of 200 nuclear power stations, 250 new coal mines, 150 coal-fired plants, 30 new refineries, and 20 new synthetic fuel plants, all large-scale.

This strategy presupposed that the conventional way of providing for energy needs that had seemed to work in the past—combining government support and regulation with industrial initiative—could be made to work again. The prospect was particularly appealing because this proposal did not call for radically lowered expectations but instead promised an economic boom resulting in an ability to rely on domestic resources. The trouble was that it rested on highly optimistic projections of what could be accomplished in a short time and on equally optimistic estimates of what such massive efforts would cost—not only in economic terms but also in terms of the social and political costs that would be entailed by efforts to override concern for preserving strict environmental and safety standards.

As the sense of crisis deepened, a very different view gathered increasing support. Articulate critics charged that the effort to increase supply to meet ever-growing expectations had brought about the energy problem and was hardly an adequate prescription for dealing with it at a time when conventional resources were being rapidly depleted. Instead, the critics called for a policy aimed at curbing growth in energy consumption—one that would be sensitive to environmental considerations and appropriate to an "era of limits," even if it did not satisfy all expectations of continuous improvements in material standards of living.[34]

This point of view had considerable influence in shaping the energy program advanced by President Jimmy Carter. In declaring the energy problem "the moral equivalent of war," he used a formula designed both to arouse public support and to justify a departure from previous policies. In the first place, the Carter "National Energy Plan" stressed conservation. The assumption behind it was that "the most effective way to 'produce' new energy was by saving it."[35] The plan also rested on the belief that in the long run it was essential to move from nearly exclusive reliance on fossil fuels to a much greater reliance on "renewables."

In foreign policy, the Carter program also reflected a rejection of the counsel of those who saw the energy crisis as a challenge to American power, which would require vigorous assertion of that power. The Carter program did not aim at breaking the power of OPEC either by threatening or using military force or by forming a countervailing alliance of consumer nations, or by using any of the other methods that were being suggested for that purpose. Instead, its aim was to hold OPEC in check by negotiating emergency oil-sharing agreements with America's allies, by building a strategic petroleum reserve, and by agreeing with other user nations to set targets for achieving lower imports of oil. The policy also included an effort to co-opt the leading Arab oil-producing state, Saudi Arabia, into a closer alliance with the western nations.

This program, although it was by no means radical enough to satisfy the harshest critics on the political left, combined overtones of the New Deal (with its emphasis on comprehensive planning and on mandatory conservation measures) with a recognition of the new concerns for a more frugal way of life, protection of the environment, and the maintenance of peaceful relations with developing countries, which were important to certain elements of the Democratic electoral constituency.

Although the full Carter program was not enacted into law, some of the most important elements were put in place by a combination of legislative and executive action. As a result, steps were taken to bring U.S. oil pricing up to world levels, to control the price of newly discovered natural gas, to impose an excise tax on newly discovered oil, to mandate conservation (by stipulating improvements in average automobile fleet mileage, establishing a national 55-mile-per-hour speed limit, and prescribing thermostat settings in federal buildings). Great emphasis was placed on the development of alternative energy sources, notably solar sources, and a Synthetic Fuels Corporation was created to channel public investment into private efforts to produce liquid fuels, as well as gas from unconventional sources. Some of the Carter proposals were rejected. The proposal for an Energy Mobilization Board was rejected by Congress, as was a request for standby authority to impose a gasoline tax. Certain of the other proposals were considerably modified, such as the windfall profits tax, which was originally intended to apply to all domestic oil, whether already or newly discovered.

Enough of the Carter program was adopted, however, to make it clear that there was considerable public recognition that the energy problem was not just a temporary emergency caused by private greed or government mismanagement, but that it reflected a long-term exhaustion of relatively inexpensive forms of fossil energy. It was also an indication of public acceptance of a basic strategy for dealing with the problem by making more efficient use of domestic resources and preparing for an eventual transition to greater reliance on renewable sources. The Carter approach to nuclear power reflected the public ambivalence on that subject. Nuclear power was not to be abandoned, but neither would it receive the same emphasis it was to have under the Nixon-Ford proposals. Safety standards would not be relaxed, more effort would be put into finding an acceptable way of storing nuclear wastes, and the development of the breeder reactor and the export of reprocessing facilities would be foregone in an effort to curb the threat of the international proliferation of nuclear weapons. The importance of developing and implementing a comprehensive strategy was recognized in the creation of the new Department of Energy.

The results of the 1980 election were interpreted by the victor, Ronald Reagan, to indicate that a majority of the voters agreed with his contention that freedom and prosperity would best be served by curtailing rather than continuing to expand the role of the federal government. Accordingly, he pledged to deal with the energy problem by encouraging private initiatives to discover and make available more sources of energy of all types. The new Energy Department was to be dismantled, and the burden of over regulation lifted from all segments of the energy industry, including the nuclear power sector. Oil prices were fully decontrolled ahead of schedule, federal lands were opened to drilling, and offshore oil and gas exploration was encouraged. Funds for the Synthetic Fuels Corporation were sharply curtailed. As oil consumption dropped and prices for oil declined, the Reagan administration took some of the credit, noting that market forces had been far more effective than government intervention. (It was not yet clear, however, to what extent these results were a temporary by-product of the global recession, and in any case, some of the decline in consumption must be attributed to mandatory conservation measures.) Although the Reagan program has yet to be fully implemented, it is already abundantly clear that the thrust of the program is closer to that advanced by presidents Nixon and Ford. Although it does not aim to make the nation independent of foreign supplies, it puts the same emphasis as earlier Republican programs on the effort to satisfy expected increases in demand by encouraging private industry to develop new supplies. To the extent that conservation is to occur, it will be as a result of higher prices rather than mandatory requirement, or as a result of initiatives on the part of state and local governments and public utility commissions.

Clearly this debate over energy policy is not simply a reflection of the economic interests of the different groups and regions, or of a rational response to new developments. It has also been influenced by considerations of political ideology. Republicans have addressed the energy problem as Republicans tend to address all other socioeconomic problems, emphasizing the need to maintain economic growth by stimulating market forces and curbing the role of government and bureaucracy. Democrats tend to address the same problem by emphasizing the need for planning and government initiative, for protecting the disadvantaged from the hardships inflicted by the play of market forces, and—lately—for adjusting expectations of continued improvements in standards of living and consumer choice to the growing constraints upon the possibility of economic growth.

The debate over energy policy is therefore only in part a debate about external realities and technical questions. It is, at the same time, a debate about the principles by which policy should be guided. There is an important difference, however, between relying on ideologies to define these principles, and relying on fresh reflection. Ideologies are by definition simplifications of complex ideas. Often, they have been formulated in earlier times, when conditions were quite different. Sometimes, they contain internal contradictions, which become evident only when they are critically examined. And more often than not, they can be invoked to rationalize or disguise a special interest by making it appear to be associated with some lofty principle. To appreciate more fully what is at stake in the debate over energy policy, it is therefore necessary to set ideologies aside and to try to identify and analyze the values in question and their implications for policy, as well as the dilemmas that arise in the effort to apply them in practice.

4 Values in Conflict

The industrial Revolution was initially powered by water and fuelwood, but its expansion was dependent on the use of coal. The direction of twentieth century growth in industrial nations has been largely determined by the availability of cheap oil. Societies of the twenty-first century will be shaped to a significant degree by the energy technologies that are chosen in the next few decades to replace oil.

We are not literally running out of energy, but the era of cheap energy is over. The rising economic, environmental, and human costs of energy decisions will affect almost every phase of our lives. With escalating energy prices, major shifts in national patterns of expenditure and investment are occurring. Such a transition provides an unusual opportunity to ask about our priorities. What kinds of things do we really value? Whose needs are most important? Who gets the benefits and who bears the risks of particular energy choices? Who should make these decisions? How can social and environmental costs as well as direct economic costs enter into decision-making processes?

Technical and economic considerations are of course crucial in policy analysis; they place constraints on what is possible and practical. But ethical principles are relevant in deciding what is desirable. If our social goals influence the directions of research and development they will affect future technological feasibility and comparative costs. In this chapter, three values are examined: individual freedom, distributive equity, and economic growth. Other values—health and safety, environmental preservation, world peace, and long-term sustainability—are treated in the next chapter. In each case, the basis of the value is traced in our national heritage, religious traditions, or philosophical principles, and applications in particular energy policy decisions are explored. These seven values have been selected for discussion because they are both important in the American heritage and relevant to energy policy choices.

A value is a general characteristic of an object or state of affairs that a person views with favor, believes is beneficial, and is disposed to act to promote.[1] To hold a value is to have a favorable attitude towards its realization; in this respect, values resemble preferences or desires. However, subscription to a value also includes beliefs about benefits or moral obligations that can be used to justify or defend it or recommend it to others; such beliefs are open to rational reflection and discussion, as individual preferences are not. When a choice is defended in terms of values, reasons are given and general characteristics and principles are invoked. The values that people hold can be studied empirically by social scientists, using verbal testimony, data on behavior, cultural expression, interviews, and public opinion surveys.[2] But values are also of interest to humanists, since they include beliefs about benefits and obligations. The philosopher, in particular, is concerned about the clarity of the concepts and the universality and consistency of the principles invoked and their relation to other beliefs. These chapters include some reference to the values actually held by Americans today, but their main concern is the analysis of concepts and ethical principles and their application to energy policy.

The relation between a particular value and a particular policy option is seldom simple. A decision may have diverse consequences, which affect the realization of a value in contradictory ways. Postponement of the deployment of breeder reactors (that is, reactors that convert nonfissionable into fissionable material) would probably retard the proliferation of nuclear weapons and thereby reduce the chances of nuclear war. At the same time it might prolong international competition for dwindling oil supplies and increase the prospects of oil wars, which could escalate into nuclear conflict. In addition, a policy that enhances one value may be inimical to another value. Many measures contributing to distributive equity or environmental preservation, for instance, involve some restriction of freedom or some cost in economic growth.

Energy policy disagreements arise partly from differences in the relative weight attached to diverse values such as freedom and equity. There are also differences in the estimation of the ways in which a particular policy will lead to the realization of a set of values. Advocates of a policy may be so preoccupied with the achievement of one desired consequence that they fail to note other unintended consequences. The prediction of consequences is partly an empirical question, and some differences can be resolved by more careful analysis of factual evidence. But prediction is always based on assumptions and simplifying models of the world, about which there may be disagreement. Moreover, there are areas of considerable scientific uncertainty—about health and safety, future economic costs, and future technological progress, for example—in which the professional, institutional, and personal biases of experts may influence conclusions in divergent ways. The separation of fact and value in policy issues is easy in principle but very difficult in practice.

In looking to the future, three time scales for energy decisions can be distinguished:

> 1. The greatest short-term danger during the next decade is the threat to world peace from conflicts over oil. The United States, Europe, and Japan are highly vulnerable because of their heavy dependence on oil imports from areas of the Middle East notable for their political instability. Possible short-term U.S. responses include stronger conservation measures, attempts to increase domestic oil production, and the creation of a strategic petroleum reserve; there are also a few situations in which coal could replace oil.
>
> 2. In the intermediate term (ten to 25 years), coal and nuclear energy are the main supply options. Coal can be used to generate electricity and to make synthetic oil and gas; U.S. reserves are abundant, but expanded use entails substantial risks to health, environment, and perhaps global climate. The use of nuclear energy to generate electricity could be expanded if uncertainties about radioactive waste disposal are resolved and public apprehension about reactor accidents is allayed. Solar energy and conservation could also make major contributions in this period, though there is wide disagreement about their social and political consequences.
>
> 3. The long-term challenge (beyond 25 years) is the transition to sustainable sources. The likely alternatives are solar energy, breeder reactors, and fusion (if it proves practical). Since it takes a decade or two for new equipment to be developed and widely deployed, plans for intermediate– and long–term technologies have to be started now, and they will compete for funds in the short term. The value conflicts examined below must therefore be considered in relation to all three time scales.

INDIVIDUAL FREEDOM

One consideration in the evaluation of any energy policy is the extent to which it restricts or extends individual freedom. But freedom has many forms, which may be affected in diverse ways by a policy decision. A particular form of freedom can be expressed as a relation between an agent, a constraint, and an activity. Explicitly or implicitly, it has a triadic structure: x is free *from* y *to* do z. People have particular kinds of constraints and activities in mind when they defend freedom. Sometimes they emphasize the absence of a constraint, and sometimes they emphasize the opportunity for an activity they deem important.

The negative side of freedom is the absence of external constraints, freedom *from* coercion or direct interference imposed by other persons or institutions. Locke and the early British tradition of libertarian political philosophy interpreted freedom primarily as the absence of interference by other individuals or by the state. They were concerned to protect the individual against abuses of the power of government; they wanted to allow the maximum scope for individual initiative in economic affairs and the use of private property. This view was influential among the authors of the U.S. Constitution and was reinforced by the American experience of the frontier, abundant resources, and the assumption that this was a land of unlimited opportunities for everyone. It seemed that a person free of human constraints could pursue the mastery of nature without interfering with other persons.

The positive side of freedom is the presence of opportunities for choice. Freedom *to* choose among genuine alternatives requires a range of real options and the power to act to further the alternative chosen. Even in the absence of external constraints, unequal power results in unequal opportunity for choice. Some degree of personal autonomy is an essential component of freedom. Many of the conditions for the exercise of choice are internal. People vary widely in their awareness of alternatives, ability to make deliberate choices, and personal initiative and self-direction. But in dealing with public policy, we are concerned mainly about the external conditions, the social structures within which people can have some control over their own futures.[3]

The negative and the positive sides of freedom are inescapably related in any social order. If we try to minimize external constraints while there are great inequalities of economic power, the weak will have little protection from domination by the strong. In a complex society, the actions of one person can greatly affect the choices open to other people. Limitations on the actions of some persons are necessary if other persons are to be able to exercise choice. Positive freedom to achieve desired outcomes exists only within an orderly society; we accept traffic lights that are sometimes red because we can go without interference when they are green. The state is an instrument of order and law, but it is also an instrument of freedom when it restricts some actions in order to make other actions possible.

Political freedom, too, has both negative and positive aspects. On the negative side are limits to the powers of government, such as censorship and arbitrary arrest. On the positive side are institutions of political self-determination, democratic forms of government whereby each citizen can have a voice in decision-making processes. Civil liberties, such as freedom of speech, assembly, and the press, can be defended both as basic human rights and as preconditions of democracy.

The American tradition has insisted on the limitation of governmental powers in order to protect the individual from too much interference by the state. But the right of governments to intervene to protect health, safety, and welfare has been expanding to include ever wider areas as the uses of private property have had more far-reaching public consequences. Such common resources as air and water can only be protected by collective action through regulations or economic incentives. In other cases, governmental powers were expanded to protect citizens from the growing power of private institutions such as industrial corporations and labor unions.

The forms of freedom that are relevant to energy policy can all be understood as *ways of participating in decisions that affect our lives.* One form is individual choice in a free-market economy, with minimum government interference consistent with public safety and welfare. Another is participation in the political processes of democratic decision making at all levels, through which governments are accountable to citizens. Let us consider four forms of freedom in recent energy policy debates: (1) the consumer's freedom *from* government interference *to* buy and sell in a free market (free enterprise), (2) the citizen's freedom *from* authoritarian control *to* have a voice in decision-making processes (political democracy), (3) the citizen's freedom *from* government surveillance *to* speak and associate freely (civil liberties), (4) the individual's freedom *from* the pervasive power of large organizations *to* exercise local control over productive activity (decentralized ownership).

Consumer Choice and the Free Market

Consumers have more choices in the marketplace when adequate energy supply is assured. As a nation, assured supply gives us greater control of our own destiny and prevents economic disruptions beyond our control. The auto has enhanced the mobility, job flexibility, and personal freedom of most Americans (except the poorest, the elderly, and the handicapped); smaller, more efficient cars can preserve these benefits, provided enough liquid fuels are available. By contrast, energy shortages would lead to increased government regulation, and extreme shortages might lead to social unrest and perhaps authoritarian measures and the curtailment of civil liberties in the interest of restoring order.[4]

Market mechanisms ordinarily provide an automatic adjustment to resource scarcities, without government intervention. When a resource becomes scarce its price rises, thereby reducing consumer demand and increasing the supply (by encouraging the search for new reserves, extraction technologies, and substitute materials). Economic efficiency as well as individual freedom is served by maximum reliance on the free market and minimum reliance on government regulation—which restricts choices and usually results in less efficient allocation along with additional administrative costs.

The market has a number of limitations, however, which can only be corrected by public action through political processes. Costs in environmental damage or human health are externalities, which can be internalized only through regulatory standards or taxes and subsidies. Dependence on imported oil is a threat to national security, which can be reduced only by collective action. Market competition is far from perfect because of cartels, monopolies, and price agreements. Market decisions reflect a short-term view because future costs and benefits are heavily discounted. There are thus many aspects of energy policy for which political as well as economic institutions are crucial.

When government action is needed, there is a spectrum of possible measures from voluntary to coercive. In the case of conservation, for example, educational programs can raise public awareness and encourage energy-saving practices. But people are reluctant to make sacrifices that are not shared by others, and many forms of conservation cannot be practiced by separate individuals. There is a minimum of coercion in public subsidies through tax credits, research funds, or federal grants. Taxes on fuels provide economic incentives for conservation but put a heavy burden on the poor unless there are provisions for rebates. Mandatory efficiency standards, such as auto mileage requirements, are politically acceptable because they apply coercion to a few companies rather than to individual citizens and because technical changes are often easier to bring about than behavioral changes. Rationing is more coercive of individuals and involves greater personal inconvenience, additional bureaucracy, and the temptations of black-market abuse.[5] In general there is a tradeoff between effectiveness and freedom here, and the more coercive measures can be justified only under emergency conditions. Coercion is more widely accepted if there is a perceived emergency and if policies are democratically adopted (a form of "mutual coercion").

Democracy and Expertise

We have identified freedom with participation in the decisions that affect our lives, whether through economic choices in the marketplace or political choices in a democracy. But political participation is difficult when both citizens and legislators feel incompetent to deal with complex technical issues such as those in energy policy. Yet policy decisions should not be left to technical experts alone, since they involve value judgments as well as scientific judgments. Who should decide? How can democracy be reconciled with the need for expertise?

One answer lies in a more informed public. Elected representatives are responsive to strong expressions of public opinion; the level of citizen awareness of technical issues can be improved through educational institutions and the media. Even so, we cannot expect any easy consensus to emerge.

since there are such conflicting interests and differences in value priorities in our society. Moreover, particular groups can sway public opinion by purchasing media time or using media coverage to promote their own narrow goals. Nevertheless, an informed electorate and extensive public debate are essential elements of democratic self-government. Citizens have a voice in setting the agenda of issues for legislative action.

Another answer lies in better scientific advice to legislatures. In the past, Congress had little scientific expertise of its own, and the experts on which it relied were drawn mainly from industry or government agencies, which were often interested in promoting particular technologies. During the 1970s an increasing number of scientists were added to the staff of congressional committees, and the Office of Technology Assessment was established to provide independent, balanced studies of policy options related to technology. Congressional hearings have been drawing from a wider range of expert witnesses, including scientists from universities and from environmental, consumer, and other public interest groups.[6]

Federal agencies are accountable to the public primarily through officials appointed by an elected president, as well as through congressional oversight and judicial review. But there are also channels for direct participation by citizens and independent experts, which can enhance the accountability and openness of agency decisions. Many legislative acts and agency regulations require public hearings in connection with rule making, standard setting, and the issuing of licenses and permits. In other cases, drafts of agency documents must be circulated for comment by industry, public interest groups, and other agencies before their final formulation.

In the case of nuclear energy in particular, there has been strong public demand for greater openness in decision making. Until the 1970s, decisions were made largely by agency experts, working closely with industry and a congressional committee dedicated to the promotion of nuclear technology. Public access to information was often severely limited, and in some instances staff dissent was suppressed. Nuclear policy was made on almost exclusively technical grounds, with little consideration of social and political issues or tradeoffs among conflicting values. Public hearings on plant sites occurred late in the decision process after the safety analysis report had been approved; government and industry were by then allies against citizens who raised questions about safety. Citizens felt powerless in the face of huge industries and a massive bureaucracy.[7]

During the 1970s, there was much more debate about nuclear energy in Congress and there has been an expansion of opportunities for public influence over agency decisions. Public apprehension about health and safety has led to more cautious regulatory policies, and licensing has been subject to long delays. The human errors and equipment failures of Three Mile Island, and the days of confusion following the accident, reinforced public skepticism about reassuring statements by experts—even though monitoring equipment and training and safety procedures will be improved as a result of the accident. On the other hand, proponents of nuclear power feel that opponents have been given too many opportunities to obstruct agency actions through hearings and court challenges. Once a basic policy issue has been decided by Congress, they hold, it should not be raised again in each rule-making or siting decision.[8]

There are two main objections to public participation in agency decisions. One is that the citizens who take part are not representative of the public. Environmental activists, for example, are predominantly middle-class, well-educated professionals. Intervenors may have a hidden political agenda that is only remotely related to the technical or procedural issues under discussion. Perhaps greater efforts should be made to encourage participation by a wider range of population segments and experts from a variety of institutions. The second objection is that public involvement results in delay and obstruction. In some cases a small group of local citizens has blocked a project from which a wider public might benefit. Special interest groups, unwilling to compromise for the common good, have occasionally exercised a virtual veto power. Clearly the right of a minority to be heard does not imply that its view should prevail. Once a decision is made at one level, it should not be made again at various other levels.[9]

If its limited role is recognized, public participation can enhance the openness of decisions, the accountability of agency officials, and the vitality of public debate, without paralyzing all action.[10] In addition, the acceptance of risks in society should be as voluntary and as consistent with "informed consent" as can reasonably be achieved. Major technological decisions inevitably impose involuntary risks, but there should be opportunities for people exposed to risks to be heard. Nuclear energy has been most rapidly deployed in nations such as Russia and France in which centralized government agencies have strong powers over technological development and plant siting; deployment has been slower in nations such as the United States, Sweden, and West Germany in which legislatures, courts, or regional authorities can more readily challenge administrative agencies.[11] Our goal should be to seek a balance between administrative efficiency and democratic participation in such decisions. Participation should be designed not to raise again issues that have been settled in representative legislatures, but to encourage the accountability of agencies by exposing their actions to public scrutiny.

Plutonium and Civil Liberties

Nuclear weapons can be made from current light-water reactor fuels, but only by uranium enrichment or extraction from spent fuels, both of which require very complex equipment. If breeders and reprocessing are deployed in the United States, however, there would be large quantities of plutonium in circulation, and it would be difficult to prevent the diversion or theft of the small quantity needed to make a nuclear bomb. Plutonium would be a tempting target for terrorist groups because it can be handled with relative ease, and a nuclear device would constitute a potent blackmail threat. Security against theft would require tighter plant security and more extensive employee loyalty investigations than currently practiced. Some critics envisage the wide use of infiltrators and wiretapping for the covert surveillance of radical groups who might attempt nuclear sabotage or the diversion of nuclear weapons material to foreign organizations such as the Palestine Liberation Organization (PLO). There have already been cases in which all employees in a plant were required to take lie detector tests about their off-plant associations, and dossiers have been gathered on nuclear protesters. If plutonium were stolen, extreme efforts at search and recovery would undoubtedly be undertaken.[12]

Some commentators see such security measures as a step toward a police state. It is true that secrecy in domestic intelligence and surveillance always involves government powers that can be abused. But as Willrich and Taylor point out, plutonium management would not require levels of security greater than we have accepted elsewhere in our society.[13] Guidelines to limit abuses of secrecy would be important here as in other intelligence operations, and greater accountability to Congress should be sought. Under conditions of social stability, tighter security could be provided without threatening civil liberties, though amid social disruption and violent conflict the protection of plutonium (and of nuclear weapons themselves) could serve as an excuse for more repressive measures.

Participation and Decentralization

Advocates of small-scale technology are concerned about another form of freedom, the opportunity for individuals to have a more direct voice in decisions about production. It is difficult for people to influence decisions about large-scale technologies, either through the marketplace by exercising the consumer's economic freedom, or through democratic processes by exercising the citizen's political freedom. Nuclear plants are inescapably large-scale and require huge capital investments; they can only be owned by large corporations, as in the United States, or by governments, as in most other countries. By contrast, many (though not all) forms of solar equipment can be owned by individuals and communities, which would encourage local self-reliance and counteract the trend toward the concentration of economic and political power. The debate between proponents of centralized and decentralized technologies can be seen as a new version of the historic conflict between Hamiltonian and Jeffersonian visions of America's future, outlined in the previous chapter.

Solar heating offers scope for small businesses in production, installation, and repair of equipment. Community-level systems, including the cogeneration of heat and electricity, could be run as cooperatives, small companies, or municipal utilities. Amory Lovins suggests that solar technologies will be based on sophisticated principles but will be relatively understandable and simple to use, reducing our dependence on experts. Smaller systems will not be subject to large-scale accidents and, therefore, will not require extensive government regulation or tight security. Solar sources are diverse and can be matched to end-use and temperature needs. The costs of some solar technologies are still quite high, but in many cases they are falling, whereas the costs of most other energy sources are rising.[14]

Critics reply that total decentralization would be neither efficient nor socially desirable. Solar components will have to be mass-produced to be cheap enough to be widely adopted. Like autos, solar equipment can be individually owned and decentralized in use but can be more cheaply produced with the economies of scale possible in large factories. Moreover, with the existing distribution of economic power it is likely that many forms of solar energy will be controlled by large companies. The social regulation of such companies may be difficult, but it is not impossible. State utility commissions, which are less centralized than the federal government, have authority to set utility rates because a utility usually has a monopoly in its service area. Even if local self-sufficiency were possible it might not be equitable; an urban ghetto or a rural village would end up with much poorer electric service than an affluent suburb. By stressing local self-reliance, decentralists have neglected the importance of social integration and cooperation for the common good. Most Americans are more interested in the cost and convenience of energy than in local control; they are willing to pay someone else to deliver energy to them.[15]

Moreover, the correlations between energy systems and social structures are loose and ambiguous. Solar enthusiasts and nuclear enthusiasts sometimes share the assumption that the right technology, the "technical fix," will solve our social problems. But the social and institutional context in which a technology is deployed is often crucial in determining its ultimate social consequences. Yet we must also acknowledge that some technologies have distinctive potentialities, which can be supported by deliberate social policies. The relationships between scale, efficiency, and equity have to be examined separately for each part of an energy system if we are to understand the tradeoffs between local participation and other values. To date there has been little empirical research on the social impacts of alternative energy systems.

A selective and discriminating use of decentralized technologies would offer many advantages. Cogeneration of heat and electricity can substantially improve efficiency; the additional fuel required for cogeneration is about half that required by the most efficient single-purpose utility plant.[16] Cogeneration can be used in relatively large industrial applications, but also in community-level systems. District heating, integrated systems, and the recovery of energy from urban sewage, solid wastes, and agricultural residues are best carried out locally. Again, dispersed production of electricity tends to reduce transmission and distribution costs, which constitute half of the consumer's electric bill. But the electric grid offers convenience and reliability in service, and it allows power to be transferred between regions with different peak-load hours or unused capacity. Local sources should be tied into networks wherever possible. Large blocks of electricity for heavy industry and urban areas will probably require central generation. Some photovoltaic installations and some biomass production can be locally controlled, but competition for scarce land may make large solar arrays in remote areas or large plantations of plants or trees for fuel desirable.[17]

The hard and soft paths do not seem to be mutually exclusive, as Lovins claims. A mix of small and large systems, adapted to varied tasks and conditions, appears preferable to either type alone. But a case can be made for the assertion that such a mix will require a deliberate effort to develop the largely untapped potentialities of smaller systems. In the past we have subsidized large-scale technologies, including large hydroelectric dams. We heavily supported nuclear development through research funding, insurance subsidy (under the Price-Anderson Act), and waste disposal and decommissioning costs (which are likely to exceed the funds that industry has set aside for them). Past investment in large systems adds to the momentum for their perpetuation, whereas the constituencies for the small are relatively weak and diffuse. There will continue to be cases in which we will want to subsidize research on large systems, such as synthetic fuel and fusion technologies. But several recent studies suggest that decentralized solar sources merit substantial subsidy in the light of their long-range potential.

Each of the forms of freedom considered above is a significant mode of participation in the decisions that affect one's life. Each has roots in the nation's heritage and is of special concern to a segment of the American public. Proponents of the hard path tend to rely heavily on economic forces in the free market. Proponents of the soft path stress the potentialities of decentralized ownership and local control. But it is political democracy which is perhaps the crucial form of freedom today. Political decisions set limits to the operation of economic forces and provide a national context for what can be done locally. While it does not appear likely that civil liberties would be jeopardized by new energy technologies, the participation of citizens in energy policy decisions remains a difficult and important issue. The main political choices must be made by an informed public through the channels of representative government, but public input in agency decisions, along with legislative and judicial review, can encourage greater accountability in federal and state bureaucracies.

DISTRIBUTIVE EQUITY

In all energy policy decisions, the question may be asked whether the distribution of costs, risks, and benefits is equitable. An inequity may be defined as an unfair inequality in the social distribution of goods. An inequality is not inequitable unless it is unfair or unjustified in relation to its causes or consequences. The term equity can be used interchangeably with distributive justice, which should be distinguished from procedural justice (equality before the law, fair trial, impartiality, due process, and so on).

The first principle of equity is that those who receive the benefits of an action should insofar as possible bear its costs and risks, or compensate those who do. It is unfair for one person to benefit from an action that subjects other persons to uncompensated costs and risks. Unpaid indirect costs or "externalities" lead to an inefficient allocation of resources; air, water, land and other resources will be overused if market costs give misleading signals because they do not reflect true social costs. Apart from misallocation, however, there is an issue of fairness if a person in one location, occupation, or generation receives the benefits of energy production for which other persons have some of the burdens (costs and risks). Measures designed to achieve a closer correlation of benefits and burdens are considered below.

The second question is more problematic. When are inequalities in the distribution of benefits fair? Unequal rewards may be considered fair recompense for differences in the past contribution individuals have made to society. Some people have special needs or handicaps that justify special treatment; unequal treatment may be aimed at providing them with more equal opportunities in life. Inequalities in income can be defended as incentives for work and greater productivity from which everyone will benefit in the long run, though it is debatable whether inequalities of the magnitude that exist in the United States are necessary to encourage productivity. Clearly there are tradeoffs between equity and efficiency. Okun suggests that transfer payments from rich to poor are like the transfer of water in a leaky bucket. People vary considerably in the amount of leakage they are willing to accept for the sake of greater equality.[18]

As a nation we have tried to ensure that all persons have access to the resources we consider necessary to fulfill the minimum conditions for human life and dignity. We have instituted entitlement programs with respect to education, health care, food, housing, and old age security, for example. We have recently become more aware of the tax burdens and the administrative and social costs that these programs entail. Some people have argued that energy, like food, is a necessity for survival, and that a basic allocation should be provided through energy stamps or "lifeline" electricity rates; however the losses in efficiency and in incentives to conservation would be high. Others have advocated tax rebates or subsidies to mitigate the impact of rising energy prices on low-income families. Still others have argued that we should rely on progressive income tax policies to reduce the inequalities in income, which lead to inequalities in access to particular goods and services such as energy. We must examine each of these alternatives.

The most influential contemporary discussion of the relation between justice and equality has been given by the philosopher John Rawls. He has argued that principles of justice can be derived by imagining a social contract between persons who do not know in what social position or generation they will live. The hypothetical contractors, he maintains, would allow only those social and economic inequalities that maximize the benefits to the least advantaged, since any of the contractors might end up in the worst-off position. Impartiality in the formulation of the rules of the social order would be achieved because no one would know what his or her status in society will be. Each person would want other persons to be treated as he or she would want to be treated in their position.[19]

In Rawls's view, inequalities in the distribution of primary social goods (income, wealth, power, and self-respect) are just only if they result in the greatest benefit for the least advantaged. He suggests that if the least advantaged benefit, it is likely that most other social groups will benefit also, but the criterion is not the total social gain but the consequences for the worst-off. Rawls has been attacked from the right for being too egalitarian and for failing to provide adequate protection for property rights and for wealth acquired by legitimate means. He has also been attacked for not being egalitarian enough; critics on the left point out that economic inequalities, once established, perpetuate themselves and grow with time, partly because economic power is a source of political power in our society.[20]

Concern for the plight of the poor has been prominent in the teachings of the main American religious traditions. Since the days of the Exodus from Egypt, the biblical God was understood to be on the side of the poor and the oppressed. The disparity between the rich and the poor was harshly judged by the Hebrew prophets in the light of their belief in the fundamental equality of all persons before God. Extremes of wealth and poverty were seen as a violation of the covenant with a God of justice and righteousness, and as a violation of human relationships within the community. In Deuteronomy, laws for the protection of the poor are presented as the requirements of justice, not as individual acts of charity. Subsequent Jewish and Christian teachings have maintained that a social order is unjust if some persons live in affluence while others lack the basic necessities of life.[21]

Today, 94 percent of American adults say they believe in God, and nearly half attend church or synagogue regularly.[22] But the record of religious institutions in practicing the ideals they profess has been very mixed. Often they have been on the side of an inequitable status quo, resisting changes in the social order. To victims of deprivation they have sometimes offered only resigned acceptance and the consolation of a future life. Yet they have also nurtured prophetic leaders who have been in the forefront of social reforms—in hospitals and prisons and the abolition of slavery in the last century, for instance, or in the civil rights and antiwar movements of the 1960s. Religious groups have an excellent record of response to immediate human suffering, such as famine relief; only recently have they given comparable attention to the social structures that contribute to hunger and poverty. Recent statements by several church bodies have emphasized social justice as the most important criterion in the evaluation of energy policy.[23]

Let us examine the distributional impacts of some specific energy policies according to two criteria: (1) the correlation of burdens and benefits and (2) the impact on low-income families. Equity between generations is discussed in a later section.

Geographical Distribution of Impacts

The burning of coal to generate electricity in Ohio and Pennsylvania creates acid rain in New York and Canada. Electricity used in midwestern cities is generated with coal from western states, whic h sustain environmental damage and social disruption. "Boom town" growth will place heavy demands on housing, schools, and social services. Coal and uranium mining on Native American lands is a severe threat to tribal rights, cultural integrity, and traditional ties to the land. Again, nuclear waste repositories impose local risks for the sake of regional or national benefits. No state wants to be the site of a repository of which other states will be the main beneficiaries.[24]

When such impacts can be reduced at reasonable cost, the additional cost should be paid by the beneficiaries. Scrubbers can reduce the sulfur emissions that cause acid rain, land reclamation can reduce the damage from strip-mining, and community services can be expanded in boom-town areas. The costs would be reflected in higher electricity rates. But it also appears equitable that compensation should be provided for those damages or risks that are more difficult to avoid, such as residual land damage and health risks. Royalties or severance taxes paid to coal-producing counties or states are one form of compensation.[25] Congress should set a ceiling on such royalties to prevent producing states from exacting unreasonable payments from consuming states. A geological site rental fee for radioactive waste disposal has also been proposed.[26] It is clearly impractical to try to eliminate or compensate for all environmental and human costs, but the major geographical inequities can be mitigated by internalizing indirect costs.

Occupational versus Public Risks

Workers in deep mines are exposed to risks from mining accidents and "black lung" disease. These risks have been falling and can be reduced further by stricter enforcement of mine safety laws, but they cannot be eliminated.

Radiation risks to workers in nuclear plants are low, but the maximum exposure level allowed for an employee in a plant is 1,000 times the maximum level for a member of the public outside the plant boundary. Risks for uranium miners are higher than for nuclear plant workers but lower than for coal miners (both per man-hour and per unit of energy).[27]

It is often claimed that exposure of workers to higher risks than the public is justified because it is voluntarily accepted. However, "informed consent" for job-related risks is seldom fully informed or fully voluntary. With limited employment opportunities and little geographical or job mobility, a coal miner may have few alternatives to working in a mine. Another defense of occupational hazards is that high-risk jobs, such as the construction of tall buildings, often offer a wage premium. But additional compensation is by no means universal, and rates of compensation for comparable risks vary widely.[28]

In some cases, individual victims of occupational hazards can be identified and compensation for injury can be provided, either under workman's compensation laws or under specific legislation such as that providing payments to miners with "black lung" disease. Injury from low-level radiation, however, is difficult to prove, since it may not be evident until 30 or 40 years after exposure, and a cancer acquired on the job is indistinguishable from one arising from environmental or dietary causes. Even if the effect could be established as a small statistical increase in incidence among a population of workers, it would be impossible to identify specific victims. Some risk sharing among a group of workers can be achieved by insurance methods, whose costs should be passed on to the beneficiaries.[29]

Impacts on Low-income Families

Energy prices have risen much faster than the inflation rate and are a disproportionate burden to the poor. In 1980, low-income families spent 22 percent of their income on heating and utility bills, compared to 5 percent for middle-income families.[30] Most of these poor families live in poorly insulated houses and often rent from landlords who have little incentive to add insulation. Now it might be argued that particular inequities should be corrected through general redistributive programs, such as progressive income taxes and the welfare and social security systems. It would be impractical to try to correct individually the inequities resulting from every piece of legislation. Nevertheless there are good reasons for including in major legislation explicit measures designed to offset undesired distributive consequences. Inequities will be taken more seriously if they are examined in connection with the legislation itself. The incorporation of measures to mitigate inequities is also likely to lead to broader public support and political acceptability. This also avoids adding to the demands on a hard-pressed welfare system.[31]

The deregulation of oil and gas was intended to provide greater incentives for increased supply through new exploration, and also incentives for reduced demand through conservation. It allows domestic oil prices to come up to foreign oil prices, which are much higher than production costs. Between 1973 and 1980, the value of proved domestic oil reserves at OPEC prices went up by more than $2 trillion.[32] The windfall profits tax was designed to prevent oil companies from gaining excessive profits from OPEC decisions. Because rapid rises in oil prices are such a heavy burden to the poor, Congress in 1980 allocated 25 percent of the windfall profits tax revenues to low-income assistance. Such corrective measures seem justified when billions of dollars are at stake and when policy is influenced by national political goals such as reduced dependence on foreign oil.

Revisions of utility rate structures have also been proposed in the name of equity. In the past, lower rates have been given to large users, partly because their distribution and billing costs per kilowatt are lower, but also in order to promote more demand. Large-volume users lower the costs for everyone, but they carry less than a proportional share of base-load costs, and they have less incentive to conserve with discount rates. Level or even inverted rate structures for larger users would correct these defects, though higher costs to industry would be passed on to consumers. Another proposal calls for very low "lifeline" rates for the first block of home electricity, with steep rises in the rate for additional amounts to encourage conservation. However this would not be specifically targeted on low-income families, and small affluent families with efficient appliances would be among the beneficiaries.[33]

Congress has legislated income tax credits for all citizens for expenditures for home insulation and energy-conserving equipment, but there are also provisions for low-cost loans to low-income families. A limited program provides emergency funds to alleviate special hardship situations in home heating. Some consumer advocates have proposed that the welfare system should include fuel stamps for a minimal home heating allotment. But this would lessen incentives for conservation and would overload an already overburdened bureaucracy. Subsidy of conservation measures specifically targeted on low-income families, by contrast, would contribute both to equity and to national goals of reduced energy consumption. Such policies are always subject to inequitable abuses in practice, and low-income families are seldom the only beneficiaries of programs intended for their benefit, but with careful administration greater equity could be achieved than by relying on the free market or general income transfer measures.

Distribution among Nations

On the average, a U.S. citizen uses twice as much energy as a European and 50 times as much as a person in one of the developing countries. With 5 percent of the world's population we burn half of all gasoline used. Our oil imports drive up the price of oil desperately needed for agriculture and industrialization in the Third World. Present inequalities are the product of many causes. Some people point to past colonialism and the exploitation of raw materials in less developed countries (LDCs) by industrial nations. Others attribute inequalities in the consumption of natural resources to differences in economic wealth, technological development, and military power. Clearly the international economy is far from a free market; cartels, tariffs, quotas, barriers to labor movement, and national political goals strongly influence resource decisions.

Are principles of equity applicable across national boundaries? Rawls considers justice only within nations, which he views as essentially independent entities. But Charles Beitz has argued that Rawls' principle of distributive justice should be applied globally, since nations today are interdependent and can cooperate for common interests. A nation's welfare or survival may be dependent on natural resources located in other nations. There is extensive international investment and trade; the market for many products is global. The distribution of the benefits that arise from international interaction can only be evaluated by principles transcending national boundaries. Beitz maintains that a hypothetical group of persons who did not know to which country they would belong would establish procedures to ensure that all persons had access to the resources necessary to satisfy basic human needs. They would accept only those inequalities that maximize benefits to the least advantaged. in whatever country.[34]

Beitz recognizes that such a principle of global justice is an ideal that cannot be realized immediately, but it can set a direction for changes in national policy and for modifications in international institutions. The principle is relevant to national policies, whether or not one believes a world government is possible or desirable. It is not incompatible with concern for liberty and justice within one's own nation or within other nations. Beitz maintains that U.S. agricultural and technological assistance to LDCs is an obligation of justice, not an optional act of charity. He also discusses special drawing rights in the World Bank, preferential tariffs for LDCs, changes in the terms of trade, and other specific policies.

In addition to equity considerations, there are pragmatic reasons for concern about the growing gaps between rich and poor countries, which are a threat to world trade and economic stability. In an interdependent world, the health of the global system affects the welfare of every nation. The United States imports more than half its supplies of 20 critical minerals. It is strongly affected by balance-of-payment deficits, global inflation, and fluctuations in resource prices. The continued frustration of the hopes of LDCs could lead to political instability, violence, and disruptive activities. As nuclear weapons spread around the world, industrial nations may face threats from revolutionary movements or desperate actions by impoverished nations.

If a principle of global justice were applied to energy, higher levels of per capita energy consumption in some nations might be justified by the high levels of technology and productivity that contribute to international trade, from which persons in all nations benefit. However the huge inequalities between nations today could hardly be justified. What policies would be appropriate if we concluded that a more even distribution of energy among nations would contribute to global equity as well as international stability? Clearly conservation in the United States will decrease the disparities and will slow the rise in world oil prices. There are significant opportunities for international cooperation in the transfer of existing technologies and in the development of new ones. Indigenous research and development should also be supported; only 2.7 percent of the world's expenditures for scientific and technological research occur in LDCs, with two-thirds of the world's population.[35]

What energy sources are most promising for Third World nations? In many areas, deforestation is creating a severe shortage of fuelwood—widely used for cooking—and is also leading to soil erosion, flooding, and desertification; reforestation would thus serve multiple objectives. Some LDCs need help in exploring potential oil fields, which could reduce their imports substantially. Nuclear energy would be useful in urban and industrial areas, especially in the more advanced LDCs. Only a few regions have the demand or the networks to distribute large blocks of electricity, but smaller mass-produced reactors (200 to 450 megawatts) are being developed in Britain, Germany, and the Soviet Union for the Third World market. Though nuclear energy has prestige as a symbol of advanced technology, some development experts and Third World spokesmen have emphasized its limitations. Nuclear plants do little for transportation, agriculture, or rural development. In countries without uranium or a strong indigenous scientific community, they perpetuate dependence on industrial nations for fuel, equipment, and expertise.[36]

Solar energy is particularly appealing to LDCs with surplus labor and severe shortages of capital. Sunlight is rather evenly distributed among nations, whereas uranium, like oil and coal, is very unevenly distributed. Solar technologies are more adaptable to local cultures and offer better prospects for self-reliance and national self-determination. Solar water pumps, methane digesters, small hydroelectric dams, windmills, and solar cells could be major contributors to local employment and rural development. Cheap solar technology would be a great boon to the sun-rich Third World and would reduce dependence on other nations for fuel. U.S. solar research is thus likely to benefit other countries as well as ourselves.[37]

Overall, the United States faces major difficulties in achieving both adequate energy supplies and a more equitable distribution of burdens and benefits within the nation, but these goals appear achievable—thanks to our level of economic development and our fortunate endowment of oil, coal, uranium, and renewable energy sources. On the global scale, however, the problems of both energy supply and distributional equity seem almost overwhelming. The United States industrialized while coal and oil were cheap; LDCs are trying to industrialize while rising energy prices are creating havoc with development plans and adding to already staggering debt loads. Continuing population growth puts increasing pressure on environments that are deteriorating in many regions, and agricultural production is limited by the shortage of energy for fertilizer and irrigation. These international dimensions of the energy crisis are taken up in the later chapters of this volume.

ECONOMIC GROWTH

The American Dream has included the expectation of economic growth and ever-rising levels of consumption and production. An optimistic belief in inevitable progress and a confidence in technology as the agent of prosperity have been characteristic of our national history. Industrial growth in America was aided by cheap fuels, abundant resources, and adequate water, air, and land to absorb industrial wastes. The economic institutions of capitalism facilitated capital accumulation and reinvestment, and American ingenuity aided the growth of technology, which has been increasingly based on science. Impressive increases in productivity and gross national product (GNP) have been accompanied by rising standards of living. The average life span in the United States has doubled in the past century.

Economic growth can be defended from a philosophical viewpoint if one adopts the principle of utilitarianism, "the greatest good for the greatest number." Early utilitarians identified the good with pleasure or happiness. According to later versions, we should choose that option that maximizes the total social welfare as determined by aggregating individual welfare, subjective preferences, or perceived satisfactions. But individual welfare, preference, and satisfaction are all difficult to quantify and aggregate. Utilitarian economists concluded that people's preferences can be roughly measured by their willingness to pay. The goal of "the greatest good" is then replaced by the goal of maximizing the dollar value of all goods and services, the GNP. Cost-benefit analysis operates in the framework of utilitarian assumptions by seeking to maximize the net balance of benefits over costs in comparing particular policy options.

Since maximizing the total good might involve very unequal distribution among individuals—or even the sacrifice of a few individuals for the sake of the majority—most utilitarians introduce a principle of equity as an additional constraint, or try to show that it can be derived from the principle of the greatest good. In utilitarianism, acts (or rules governing acts) are judged entirely by their consequences; there are no acts that are right or wrong in themselves regardless of their consequences. Freedom is defended in terms of its social consequences rather than in terms of fundamental individual rights. While utilitarians often emphasize economic growth as a desirable social goal, they are usually also concerned about freedom and equity.[38]

Economic growth can also be justified from the perspective of the Western religious traditions. Rising standards of living have led to a reduction in disease, illness, poverty, human suffering, and back-breaking work. The laws of Deuteronomy and the message of the Hebrew prophets call for action to alleviate physical need, particularly among the underprivileged. In the New Testament, love is not mere sentiment but active caring for persons and response to the needs of the neighbor for food, clothing, and health. "For I was hungry and you gave me food... I was naked and you clothed me, I was sick and you visited me."[39] Subsequent Christian teachings call us to respond to human need through social institutions as well as through individual actions. Material needs are taken seriously, though they are seen in the context of the total life of persons in community. The burden of poverty and the temptations of affluence are both portrayed. Here material progress is to be judged by its contribution to interpersonal relationships and the quality of the community's life as well as to the fulfillment of individual needs, especially among the poor.[40]

In the United States, the poor have benefited more from overall economic growth than from measures aimed at distributive equity. The standard of living of low-income families has doubled in a generation, though relative inequalities changed little. Even though the rich benefit most from economic growth, many gains eventually "trickle down" to those at the bottom. The economy can be compared to a pie that is growing in size; even those with the smallest slices are better off, without any change in their relative shares.

By the early 1970s, however, a number of writers were claiming that such economic growth cannot continue because of resource and environmental constraints. *The Limits to Growth* argued that if population and industrial production continue to grow exponentially, global limits would be exceeded within a few decades. The constraints arise not from literally running out or encountering absolute limits, but from diminishing returns in the use of land, accessible ores, and pollution control technology. The only way to avoid disaster, it was claimed, is to halt industrial growth as well as population growth.[41] We may note that, if the pie is not growing, the least advantaged can gain only by changes in their relative shares. In a no-growth society it is likely that social conflicts would be intensified as each group defended its own interests. It would be a "zero-sum" game in which one person could gain only if someone else loses.

Such claims of imminent resource and environmental limits seem to give too little weight to the role of creative technological advances. New technologies can turn previously useless substances into useful resources and can continue to reduce pollution per unit of production. To be sure, we must be more aware of the human and environmental impacts of technology than we have in the past and recognize that there are likely to be delayed and unexpected effects. Growth will have to be more selective, and the mix of goods and services should shift toward human services and toward less-polluting and resource-intensive products. Products can be designed for greater efficiency, durability, and recyclability. The information technologies, such as electronic communications and microcomputers, are not resource-intensive, and technologies of recycling can reduce waste as well as resource use. The American industrial base should indeed be expanded, but it should also be modified. In the Third World, economic growth and appropriate kinds of industrialization are absolutely essential to human welfare.[42]

In the United States, economic growth would stimulate research and capital investment and thereby encourage more rapid replacement of existing equipment with less polluting and more energy-efficient equipment. By the early 1980s, economic stagnation and high rates of inflation and unemployment had led many people to give economic growth higher priority than other national goals. But economists are by no means agreed as to how to deal with recession, inflation, and unemployment simultaneously. The higher costs of energy and resources, together with world economic instabilities, are likely to lead to slower U.S. economic growth than occurred in the 1960s. The expectations of both the U.S. public and the economic analysts have been scaled down during recent years. With slower growth, there is less justification for dismissing equity issues on the assumption that growth is the chief way of helping the underprivileged. Let us assume, then, that moderate rates of economic growth are both possible and desirable for the next few decades.

Economic Growth and Energy Growth

It has sometimes been asserted that substantial energy growth should be promoted in the United States for the sake of economic growth from which everyone, including the poor, would benefit. This has been one of the strongest arguments for nuclear energy on ethical grounds.[43] During the 1950s and 1960s, electricity use doubled in each decade. It was thought that the continuation of such a growth rate would require rapid nuclear deployment and the use of breeder reactors. During the 1970s, growth rates fell drastically because of economic recession and curtailment of demand as the price of electricity rose. The reserve margin (the difference between generating capacity and peak demand) grew from 19 percent of peak demand in 1970 to 34 percent in 1979 (15 percent is considered a prudent margin).[44] Future demand projections by the Department of Energy were repeatedly revised downward.[45] New orders for nuclear reactors virtually stopped, mainly because high interest rates, regulated electricity rates, and uncertainties about future safety regulations made it very difficult for utilities to borrow the huge amounts of capital needed for nuclear plant construction. By the early 1980s, higher oil prices, more efficient autos, and other conservation measures were beginning to reduce oil consumption.

How far can energy growth fall without serious consequences for economic growth? In the past, energy consumption and GNP have indeed grown together. But this historical correlation occurred while energy prices were falling, and it will not necessarily hold amid rising prices that will motivate conservation and greater efficiency. There is a rough correlation between energy use and GNP if one compares developing nations with industrial nations, but the correlation breaks down if one makes comparisons among industrial nations. Nations with a GNP per capita close to ours use half as much energy per person.[46] Improvements in efficiency serve to decouple GNP and energy. According to the CONAES report of the National Academy of Sciences, the ratio of energy to GNP could be from a third to a half of its 1973 value. Various low-growth and even no-growth energy scenarios were projected in which GNP grows at 2 or 3 percent annually, though such estimates are dependent on assumptions about price elasticity, labor productivity, and political climate.[47]

Some conservation measures can be instituted within a few years, but others require decades because of the slow turnover of homes and industrial equipment. In the meantime, sudden and severe energy shortages would result in economic and social disruption, from which the poor would be likely to suffer most. The 1973–74 oil embargo led to job layoffs and economic recession; minority groups were laid off first and low-income families were hardest hit. Assurance of adequate energy supply for a healthy economy is a fundamental national goal. But it appears that with a combination of energy efficiency and conservation measures (including the cogeneration of heat and electricity), and a mix of coal, nuclear, and solar sources, the United States could be energy self-sufficient by the year 2000. Both energy prices and the GNP would have risen, but the percentage of the GNP spent on energy would differ little from that in 1973. The impact of energy prices on developing countries, however, will be much greater; they can save little by conservation, and substantial growth in energy supply is essential for their economic and social development.

Energy and Employment

Few events in life can threaten a person's dignity and well-being as much as continued unemployment, which also adds to the burdens on the welfare system. It has been widely assumed that energy growth promotes employment. A proposed oil refinery is likely to be opposed by environmentalists as a source of pollution but supported by labor unions as a source of jobs. The labor movement has favored energy growth and has been mainly pronuclear. In 1976, the AFL-CIO urged rapid development of nuclear power, though some of its locals disagreed. On the other hand, the United Mine Workers have called for a nuclear moratorium, and sheet metal and machinists' unions have supported conservation and solar energy as sources of new jobs. The Oil, Chemical and Atomic Workers, which includes nuclear plant workers, has advocated the reduction of occupational radiation standards to the level allowed for the public, and it opposed the Clinch River breeder reactor. Some unions have been almost exclusively interested in jobs and wages, and others have given much attention to health and safety issues.[48]

Labor has often sided with industry in opposing environmental regulations that might lead to plant closings. Even the threat of loss of jobs is a powerful political weapon, though two out of three threatened plant closings during the 1970s did not occur, and many of those that did were marginally profitable or had obsolescent equipment and would soon have had to close anyway. Moreover, environmental legislation has created far more jobs than it has endangered. During a six year period, 21,900 persons lost their jobs because of plant closings due at least in part to air and water standards, but 678,000 new jobs were created in the production and operation of pollution control equipment.[49] Environmentalists and labor unions have begun to cooperate on issues of occupational health and pollution in the workplace. They have also made common cause in emphasizing the potential for new jobs in energy conservation and solar energy.[50] The higher price of energy has resulted in some substitution of labor for energy, adding 750,000 new jobs in 1979, by one estimate.[51] A continuation of this trend would help employment, though it would lower productivity (output per working hour), and if carried too far there might be adverse effects on economic growth.

Energy and the Quality of Life

According to one survey, nuclear advocates tend to see economic growth as the central issue in energy policy, whereas nuclear opponents tend to stress the quality of life as measured by noneconomic criteria.[52] One way to measure the quality of life is to use social indicators such as literacy, life expectancy, housing statistics, crime rates, suicide rates, and environmental quality indices. By many of these criteria, U.S. well-being has declined in the last 20 years. Sweden has a GNP per capita close to that of the United States, and it uses 40 percent less energy per capita, yet it outranks us on almost all social indicators. A survey of 35 industrial nations shows little correlation of social indicators with energy use.[53] The total energy used seems less significant for national well-being than the way it is used.

The GNP is not an inclusive measure of national welfare or human well-being. It ignores goods and services that are not marketed, such as the services of persons working in their homes, or clean air and water, improvements in health, national parks, and other environmental benefits.

The pursuit of productivity has often led to neglect of the quality of work. In industrial societies today there are few meaningful work roles, and a sense of fulfillment or self-respect on the job is rare. Participation in work-related decisions and opportunity for at least some creativity and pride in the use of skills can enhance job satisfaction. The development of distinctively human potentialities requires personal responsibility, enduring human relationships, and community cohesion, which are difficult to achieve amid the impersonality and anonymity of large organizations and large cities.[54]

Noneconomic goals are important for human fulfillment, and they cannot be measured by GNP or energy consumption. But for most of the world's population such goals cannot be achieved without economic growth and energy growth. Only in an affluent society could one take for granted the satisfaction of basic human needs, which for most of humankind remain unmet. Even in the United States, there are millions whose concern for material progress is understandable and legitimate. Yet we must acknowledge that if energy policy and other national choices are dictated by economic growth alone we may find that we have jeopardized some other values. Concern for adequate energy for a healthy economy need not exclude concern for the multiple dimensions of human well-being which people have in mind when they talk about "the quality of life."

Yankelovich and Lefkowitz have summarized a series of surveys from 1971 to 1979 concerning public attitudes toward economic growth. Responses to a variety of questions indicated declining expectations concerning future growth and increasing willingness to give up some growth in consumption if it would achieve lower inflation rates and preserve past gains. "In a historic change of emphasis, the American people seems to be moving toward a provisional acceptance of a trade-off where some growth is given up in favor of economic stability and nonmaterial values." Among the new goals are self-fulfillment, personal growth, choice of lifestyles, and close human relationships—goals found in the early 1970s among young, middle-class, better-educated groups but later evident among other population segments. The authors emphasize that there are among the public very diverse views, considerable confusion, and disparities between attitudes and behavior, but they conclude that there have been significant changes:

> The research reveals a picture of Americans midway between an older post-World War II attitude of expanding horizons, a growing psychology of entitlement, unfettered optimism and unqualified confidence in technology and economic growth, and a present state of mind of lowering expectations, apprehensions about the future, mistrust in institutions, and a growing psychology of limits.... It suggests that we may be moving toward a difficult new balance between material and nonmaterial aspirations.[55]

Whether this trend will be reversed during the 1980s remains an open question. In any case, it appears that with moderate economic growth and with improved efficiency, which decouples energy growth from economic growth, we could achieve very low or even zero energy growth by the end of the century without sacrificing human welfare.

In sum, the three values considered in this chapter—individual freedom, distibutive equity, and economic growth—are often in conflict with each other and with the environmental and long-term human values discussed in the next chapter. We have argued, however, that energy policies can be designed to contribute to each of these three fundamental values. The need for expertise is not incompatible with participation by citizens in democratic decision making. The inequitable distribution of burdens and benefits from energy decisions can be at least partially mitigated by legislative measures. Economic growth will only be moderate and will have to take new directions because of higher energy costs, but neither employment nor the quality of life need be sacrificed if conservation efforts are more vigorously promoted. Specific policies aimed at these goals are presented in a later chapter.

5 Uncertain Risks

In addition to the values discussed in the previous chapter—freedom, equity, and economic growth—there are four additional values relevant to energy decisions: health and safety, environmental preservation, world peace, and long-term sustainability. Unlike the values already considered, the values treated here are threatened by uncertain risks of which we have only recently become aware.

It was not until the 1970s that the public began to take seriously the problem of air pollution and acid rain from coal, or the possibility of nuclear reactor accidents. There has been a new level of concern that a nuclear war might be triggered by an escalation of international conflicts over oil, or by nuclear weapons obtained through the diversion of material from nuclear reactors. There are long-term risks that would affect future generations, such as climate changes caused by carbon dioxide from the burning of fossil fuels and radiation from radioactive wastes. We must examine each of these types of risk and its implications for energy policy.

HEALTH AND SAFETY

Concern for health and safety derives from respect for persons and convictions about the sanctity of human life. But in allocating finite social resources, we cannot assign infinite value to an individual life. Ethical issues arise in comparing relative risks, in estimating and responding to uncertain risks, and in deciding what price should be paid to reduce risks. There are no risk-free solutions; there are risks associated with energy shortages as well as with all energy sources.

Consider the risks from coal and nuclear energy, the main alternatives for the generation of electricity in the United States for the remainder of the century. *Occupational risks* from coal are greater than those from nuclear energy. By one estimate, the operation of a 1,000-megawatt coal-fired plant is responsible for two occupational fatalities per year from mining, transportation, and plant operation, compared to 0.5 occupational fatalities from the normal operation of a comparable nuclear plant.[1]

Public health risks from coal are almost certainly greater than those from nuclear plants in routine operation. Sulfur dioxide (SO_2) from coal combustion creates sulfates, which aggravate chronic respiratory disease, but estimates of the resultant fatalities vary widely, leading to continuing controversy as to whether the cost of installing stack scrubbers is justified. On the other hand, public exposure to radiation from nuclear plants is far lower than natural background or medical x-ray radiation levels.[2] There is some uncertainty concerning radioactive waste disposal, but risks from deep repositories are likely to be very small, though of very long duration (constituting risks to future generations, which are discussed later).

Catastrophic risks are the greatest area of uncertainty. In the case of coal, the buildup of carbon dioxide (CO_2) could have catastrophic effects on global climate patterns, though probably only after many decades. In the nuclear case, the greatest catastrophe would be nuclear war from weapons proliferation, but a major reactor accident could be a regional catastrophe, and reactor safety has been a subject of intense controversy. We may note several ethical issues in the analysis of these risks.

Dealing with Uncertainty

The public is confused when competent experts give widely divergent estimates of the risks arising from particular energy technologies. Why do experts disagree? First, there is scientific uncertainty arising from limited data or inadequate theories. If SO_2 causes a slight increase in fatalities from chronic respiratory disease, or if radiation causes a minute increase in cancer fatalities, the specific victims cannot be identified among those who have the same symptoms from other causes. It is extremely difficult to obtain reliable statistical studies of low-probability events among large populations. The effects of radiation, like those of cancer-causing chemicals, may be delayed 20 to 40 years after exposure. Extrapolation from high doses to low doses is uncertain when little is known about the causal mechanisms involved.[3] Again, the CO_2 balance between atmosphere and oceans and its effects on climate are only partly understood.

Second, accidents are the product not only of equipment failures but of human errors or deliberate acts, which are difficult to predict. The Rasmussen Report on nuclear reactor safety had very little data on which to base estimates of the role of human errors, which were later to figure prominently in the Three Mile Island (TMI) accident. The report ignored problems of sabotage, terrorism, or weapons proliferation, which other studies have taken very seriously even though they cannot be quantified.[4] Plant safety also depends on social conditions and the reliability of human institutions, concerning which judgments vary widely.

Third, professional biases inevitably affect an expert's perception. Disciplinary training influences the way issues are formulated and problems are conceptualized and bounded. The time scale employed may emphasize short-term or long-term effects. Simplifying models selectively represent the aspects of the world with which one's discipline can deal. Implicit assumptions may lead to divergent inferences from the data. An environmental engineer and an ecologist, for example, may bring differing assumptions to their estimates of the environmental impact of a synthetic fuel plant.

Fourth, institutional biases influence an expert's judgment. Most people either agree with the goals of their employer when they take a job or adjust their goals to those of the organization for which they work. Every institution has a limited set of objectives, including the perpetuation of its own power. As the saying goes, "Where you stand depends on where you sit." A scientist working for the nuclear industry and one testifying for the Sierra Club may both be well informed about nuclear reactors, but their estimates of reactor safety may be based on different assumptions. Institutional biases are particularly likely to enter risk assessments that involve major scientific uncertainties or judgments about human errors and actions. People interpret events and data unconsciously in ways that support their interests.[5]

There are several ways in which the problem of conflicting expertise can be reduced. Further research can resolve many scientific controversies, though this takes time and some decisions cannot be delayed. Experts have a responsibility to warn the public of potential dangers, without either exaggerating them or minimizing them to support policy positions adopted on other grounds. They should indicate their assumptions, acknowledge the uncertainties, and state conclusions with appropriate tentativeness. Professional and institutional biases cannot be escaped, but advisory panels and legislative or regulatory hearings can draw scientists from diverse disciplines and institutional affiliations. Discussion among experts and the opportunity to cross-examine each other can often clarify the areas of agreement and disagreement that are obscured in the rhetoric of public advocacy.[6]

Even when experts agree on the range of uncertainty in data, they may disagree on the policy implications because they have differing views of the nature and the assignment of the burden of proof. One party may assume that a product or process is acceptable unless there is strong evidence that it is unsafe, while the other assumes it is unacceptable unless there is strong evidence that it is safe. In some cases, regulatory legislation assigns the burden of proof. For example, new pesticides and certain types of chemicals must be tested before marketing, and the burden of proof of safety lies with the manufacturer; but for a registered pesticide or a chemical already in use, the burden of proof of risk lies with the EPA. In other instances, such as the Reserve Mining case, courts have accepted "suspected but not completely substantiated" evidence of risk as grounds for action under precautionary legislation designed to protect the public.[7] As we shall see, assessment of energy technologies is particularly difficult because there are diverse sorts of risks, which must be balanced against even more diverse economic, social and environmental costs and benefits.

One policy response to the uncertainties in the assessment of energy risks is to try to keep the options open as long as possible. The Ford-Mitre study, for example, concludes that neither coal nor nuclear energy has a clear advantage. "The range of uncertainty in social costs is so great that the balance between coal and nuclear power could be tipped in either direction with resolution of the uncertainties." The report recommends that both options be kept open as a hedge against uncertainty and an opportunity to improve the safety of both.[8]

Catastrophic Risks

In risk-benefit analysis, a risk is defined as the probability of an event multiplied by the magnitude of its consequences. Catastrophes are low-probability high-consequence events. Studies of risk perception show that most people overestimate the probability of catastrophes and underestimate low-level chronic effects. The public seems to have a fear of large-scale accidents more than proportional to their size. People also have an emotional dread of unfamiliar hazards from new technologies, and remain complacent about equivalent risks from familiar products or activities closer to home.[9] However, there may be valid grounds for special concern for avoiding catastrophes. Large-scale disasters have a visibility and produce social disruptions greater than a numerically equivalent succession of small accidents. In addition, low-probability high-consequence events can seldom be estimated with the same accuracy as high-probability low-consequence ones, and the consequences of misestimating probabilities are more serious.[10]

It is sometimes said that complex large-scale systems such as nuclear plants are more vulnerable to disruption than coal-burning plants or smaller systems. With complex systems there are of course more things that can go wrong, though this is in large measure offset by redundancy and back-up safety systems. Complex systems do sometimes behave in ways that take people by surprise (such as the system interactions that resulted in the 1977 blackout in New York City, or the reactions that created a hydrogen bubble in the TMI reactor). Human errors can interact with equipment failures in unexpected ways. Such errors can be reduced with better instrumentation and personnel training, as well as strict safety monitoring by government and industry. But complex systems can create human overload; they make high demands on managerial competence and the ability to diagnose unforseen difficulties. Nuclear plants are also said to be vulnerable because their potential for large-scale damage would make them tempting targets for blackmail threats by terrorist groups, even if security measures were increased considerably beyond their present levels.[11]

Behind some of the current dispute over reactor safety lie divergent appraisals of the reliability of human institutions. Nuclear proponents point to the fact that no human life has been lost in a commercial reactor accident and that safety regulations have been gradually tightened. They have confidence in the dedication of industry and of government regulators to public safety. Nuclear opponents point to the fact that at TMI, unknown to the operators, the reactor core was partially exposed for 13 hours and heated to 4,000 ° F or more, so that the core became partially molten; there were two days of confusion in the control room.[12] Opponents have a greater sense of human fallibility and the tendency of all institutions to pursue their own self-interest, which only partially coincides with the public interest. There are also divergent opinions as to whether we face a world of increasing conflict, social unrest, and political instability in which terrorism is likely to increase. Estimates of reactor risks are thus strongly influenced by assumptions about the institutional context of nuclear technology.

In the face of such radical uncertainties, which make risk-benefit calculations highly speculative, some authors have urged an alternative decision rule. If the potential benefits and the best-case consequences from two options are comparable, choose the one with the most favorable worst-case. This is a *maximin* rule: maximize the minimum payoff. Choose the least catastrophic risk.[13] In worst-case scenarios for nuclear reactor accidents, the total number of fatalities is comparable to the worst disasters from dam failures and liquid natural gas explosions but far greater than any disasters in coal mines or plants.[14] Fatalities from a dam failure all occur in a short period of time. The risks from a reactor accident include (with very low probability) some prompt deaths, plus more numerous latent fatalities spread out over the subsequent 20 to 40 years as a small statistical increase in cancer deaths and genetic defects among a larger population—but still the product of a single accident. Fatalities from black lung disease or air pollutants from coal plants are not classified as catastrophes because they are not the product of discrete identifiable events.

It does not appear plausible, however, to make catastrophe avoidance the only decision principle. There are no clear criteria for the "maximum credible accident"; probability-consequence curves seldom have a sharp cutoff point. Moreover, the economic costs of avoiding accidents must be considered. At present, projected costs of the main alternative energy sources with current safety equipment are roughly comparable, but technological advances could greatly alter these estimates. In short, catastrophe avoidance should perhaps be given heavier weight in policy decisions than traditional risk-benefit analysis suggests, but it cannot be the only criterion for choosing energy technologies.

Economic Costs of Reducing Risks

A few legislative acts have mandated that safety alone is to be considered. The Delaney Amendment, for example, states that a food additive is to be banned if it is found to induce cancer in man or animal; the benefits of an additive, no matter how large, are not to be weighed against a confirmed cancer risk, no matter how small. Most legislation, however, states the dual objective of reducing risks and minimizing economic costs. The way in which these conflicting goals are to be balanced in setting standards is usually left to the regulatory agency. The costs of regulation include direct costs of compliance, governmental administrative costs, and indirect costs in loss of productivity and inflationary impact.

The 1970 Clean Air Act stated that air quality standards were to be based on health data alone, "to protect public health... allowing an adequate margin of safety." It was assumed that for each pollutant there is a threshold below which there are no hazards at all. But if there is no threshold, and if the cost of removing the pollutant increases rapidly as 100 percent removal is approached, there may be very large costs in achieving very small additional benefits in the pursuit of absolute safety. Some weighing of costs and benefits would then be desirable in setting standards. In the case of SO_2 from coal, there are considerable uncertainties in the health data, but it appears that the economic benefits in health and life expectancy from achieving the air quality standards would indeed exceed the direct economic cost.[15] But the cost is high; scrubbers add an estimated 14 percent to the consumer's electricity bill. EPA regulations introduced in 1979 require scrubbers on new plants, but not on old ones if they use low-sulfur coal. While air quality standards are supposed to be set without reference to economic costs, individual plant emission limits can take abatement costs into account.

Regulatory agencies vary greatly in their use of quantitative methods in balancing risks and costs. The Occupational Safety and Health Administration (OSHA) is mandated to set standards to protect the health of workers "to the extent feasible, on the basis of the best available evidence." In 1981 the Supreme Court ruled that in setting cotton dust standards OSHA does not have to balance the benefits to workers against the employer's cost of compliance, as long as the protective technology is available. On the other hand, the Nuclear Regulatory Commission requires equipment to reduce radiation if it costs less than $1,000 per person-rem per year of exposure reduction. (A "rem" is a base measure of biological effect of radiation, i.e., *r*oentgen *e*quivalent in *m*an.) This is a cautious figure, which probably attributes a far higher value to a human life than most safety expenditures. But it allows some flexibility in administration; emission standards are not as stringent for older plants for which control equipment would be more costly. (These flexible emission standards must of course operate within the limits for maximum radiation exposure for workers and the public mentioned in the previous chapter.)[16]

In a time of inflation there is renewed interest in cost-benefit analysis (CBA) as a way of preventing "costly and excessive regulation." The dollar value of a human life is usually assigned by projecting future earnings, typically of the order of $300,000. Industry estimates of costs are often exaggerated, while estimates of risks by environmental, consumer, or labor groups are also likely to be exaggerated. Analyses should draw from a variety of experts, and studies carried out in diverse institutional contexts should be compared. Quantitative comparisons of similar risks are desirable to avoid the misallocation of limited funds; if we can save a lot more lives for the same cost, we should do so, other things being equal. CBA can be useful in well-defined decisions between specific options having similar risks and benefits, but it has severe limitations when used in broad policy decisions involving very diverse risks and benefits.

The magnitude of a risk is a scientific judgment, but the acceptability of a risk is a value judgment involving other factors besides its magnitude. Voluntary risks (in sports, autos, and smoking, for instance) are widely accepted at levels roughly 1,000 times higher than involuntary ones, such as public hazards over which a person has no control. Risks to identifiable individuals, such as a trapped miner or a critically injured child, are taken more seriously than statistical risks, such as those from carcinogens, in which specific victims cannot be identified. Moreover, CBA deals with aggregated costs, risks, and benefits, ignoring equity issues in their distribution. For example, most of the beneficiaries may be wealthy, while most of those incurring burdens may be poor.[17] The balancing of incommensurable costs and benefits should therefore be made by accountable officials and not by experts using CBA alone. Open decision-making processes in agencies together with congressional and judicial review of regulatory actions are the main means for maintaining accountability.

The indirect costs and benefits of environmental regulations are very difficult to measure. The Council on Environmental Quality estimates that the contribution of federal environmental regulations to the annual inflation rate rose to 0.5 percent in 1974, but fell to 0.2 percent by 1978 when much of the required pollution control equipment was in place.[18] Other factors, such as federal budget deficits and rising food and fuel prices, played a much larger role in inflation. Moreover, inflation is measured by the Consumer Price

Index, which does not adequately reflect the long-term social costs of pollution in public health, property damage, and reduced agricultural yields, nor the nonmarket benefits of recreation and environmental preservation. In some cases, such as water pollution, delays in abatement would result in far higher cleanup costs later or else in significant deterioration in health and agriculture. In other cases, such as nitrogen oxide emissions from autos, regulations can be postponed with no cumulative or lasting effects.

Environmental and safety regulations have also been blamed for the lower growth in productivity (output per working hour) in the 1970s. But most studies attribute from 5 percent to 15 percent (and in no case more than 20 percent) of the drop in productivity growth to such regulations.[19] The substitution of labor for more costly energy and the shift from manufacturing jobs to human services have played much larger roles in productivity changes. In any case, the direct and indirect costs of regulations have come under increasing scrutiny by the Office of Management and Budget, and in a period of recession and inflation there is an inclination to resolve any uncertain tradeoff in favor of economy rather than safety or environmental preservation.

The late 1970s and early 1980s have also seen a trend toward the use of economic incentives rather than the specification of particular technologies for pollution abatement; given greater flexibility, industry can often devise cheaper ways of achieving a desired goal. For example, instead of setting limits for each emission source in an industrial plant, EPA has introduced the "bubble concept" in which a limit is set for total emissions from the plant (as if it were covered by a bubble with a single outlet); the company may find it less costly to control some emission sources far below previous levels if other sources are less stringently controlled.[20]

Some authors suggest that even if the relaxation of environmental regulations adds to some specific risks, these would be more than compensated for by the general improvement in health and safety resulting from more rapid economic growth.[21] In the past, many risks have fallen as standards of living have risen. Whether this trend will continue amid the stresses and pollutants of high-consumption technological societies is more debatable. There is little correlation between GNP and health among industrial countries. For example, the United States has had a higher GNP per capita than any other nation, and yet 13 nations have lower infant mortality rates, and in 15 the life expectancy is higher.[22] Once general levels of sanitation, nutrition, and public health are established, the distribution of health care services within the population, along with differences in diet and psychological stress, seem to be more important factors in life expectancy than the GNP per capita. Assessment of direct and indirect risks to human life and health will continue to be an important issue in the choice of energy sources and the regulation of energy technologies.

ENVIRONMENTAL PRESERVATION

Almost all forms of energy technology have some impact on the environment. Here we examine effects on the land and its ecosystems. These ecological impacts affect human welfare indirectly, rather than directly as in the case of health and safety risks above. How do we respond to tradeoffs between energy and the environment?

Attitudes toward the environment in American history have been very mixed. In much of our past, nature was seen as an enemy to be conquered with the advance of civilization, or as a source of raw materials to be exploited for industrial growth. But there is also a long tradition of respect for nature, from Jefferson's appreciation of rural America to Thoreau's praise of wilderness experience and Muir's campaign to preserve wilderness areas. National parks were set aside for the sake of natural beauty, recreation, and habitats for wildlife. Since early in this century, the conservationists have been dedicated to the "wise use" and "scientific management" of public lands as a national resource. The preservationists wanted to preserve some areas with minimal human impact, a goal finally adopted in the Wilderness Preservation Act of 1964. By the 1960s we were more aware of the detrimental impacts of human activities on all aspects of nature. The National Environmental Policy Act of 1969 declared the goal of "the widest range of beneficial uses of the environment without degradation," and required environmental impact statements on all federal projects "significantly affecting" the quality of the environment.[23]

Stewardship of nature has been a theme in American religious traditions. To be sure, the verses in Genesis asserting that man has been given dominion over nature have sometimes been used to justify destructive practices. But the dominant biblical view is respect for nature as God's creation, valuable in itself and not simply as an instrument of human purposes. The land is said to belong ultimately to God; we are only trustees or caretakers, responsible for the welfare of the land that is entrusted to us, and accountable for our treatment of it. The Sabbath is a day of rest for the earth and other living things as well as for humanity; every seventh year the fields are to lie fallow. Many biblical passages, notably in Job and the Psalms, express appreciation and wonder in response to nature. Recent writers have sought to recover this outlook in an era of environmental degradation. Stewardship of nature is a recurrent theme in Jewish and Christian writings, but it does not have as high a priority as social justice.[24]

Ideas from ecology were a major source of new attitudes toward the environment in the 1960s. We became more aware of the vulnerability of the biosphere and the indirect repercussions of our actions. There is a wider recognition of human dependence on nonhuman nature and of the interconnectedness of the community of life. We can see wild areas in a new way as habitats for endangered species and treasuries of genetic diversity. All of these ideas have contributed to the environmental movement.

Public support for environmental protection remained strong throughout the 1970s. In a 1979 survey, 50 percent said spending on environmental programs was "too little," 15 percent said "too much," and 31 percent "about right." In a *Newsweek* survey in June, 1981, 58 percent of adults said that environmental regulations are worth the added cost to products and services, and 75 percent agreed that "it is possible to maintain strong economic growth in the U.S. and still maintain high environmental standards." But the majority was willing to accept some environmental risks for the sake of energy. Seventy-six percent favored increasing oil exploration on federal lands, and 70 percent favored enlarging the area of offshore drilling on the East and West coasts. Fifty-five percent favored "relaxing clean air requirements to permit industry to burn more coal rather than imported oil" (versus 36 percent opposed), and 48 percent favored "easing restrictions on strip mining to provide more coal" (versus 39 percent opposed).[25]

Envirionmental Impacts of Fossil Fuels

Coal has a far greater environmental impact than any other current energy option. In Appalachia, deep-mine wastes and extensive strip-mining have resulted in depressed land values, polluted streams, and the marring of areas of great natural beauty. Coal from western surface mines is cheaper and safer to mine and lower in sulfur than eastern coal, but even with reclamation measures there is considerable damage to the landscape. Environmentalists were on the whole pleased with the Surface Control and Reclamation Act of 1977, despite its limitations. It requires that reclamation efforts be sustained for five years, or ten years in arid areas. High walls at the edges of stripped areas must be back filled and the original contours restored "except where impractical." But in the Southwest, ecosystems are fragile and reclamation will compete with other demands for scarce water. Soils in the high plains can seldom be restored to their original chemical stratification, which influences the ecology. In many areas, reclaimed land can only be used for pasture, and in some cases it will take decades for plant life to be restored. The processes for making synthetic oil and gas from coal require large amounts of water, and strip-mining would be accelerated. Extraction of oil from oil shales would make heavy water demands and would produce huge quantities of residues (unless the shale can be heated underground so that the oil can be extracted directly).[26]

SO_2 from coal combustion is not only a local hazard to human health; it also forms acid rain, which falls hundreds of miles away. The harmful effects of acid rain on fish, forests, and agricultural yields are becoming evident. These environmental impacts provide additional reasons for requiring scrubbers as our use of coal expands.[27] The greatest potential risk, but also the most uncertain, is the effect of CO_2 from burning coal and oil. If present trends continue, the concentration of CO_2 in the earth's atmosphere is expected to double by the middle of the next century. But scientists are uncertain about the effect on climates. CO_2 cuts down on heat radiation (the greenhouse effect), and typical temperatures might rise as much as 3 ° C (and considerably more near the poles). If polar ice sheets melted, ocean waters would rise 200 feet, submerging many coastal cities. The alteration of air currents and rainfall could drastically affect agricultural patterns. These changes would probably occur slowly enough that we would have some warning and could perhaps act to mitigate the disaster. But the CO_2 problem is likely to curtail coal use long before coal reserves are exhausted.[28]

The conflict between environmental preservation and the need for energy became more acute during the 1970s, and environmentalists lost in several protracted disputes. The Alaska Pipeline Bill (1973) exempted environmental impact studies of the proposed pipeline from further scrutiny or litigation; the Senate vote was exactly divided, and Vice-President Agnew cast the deciding vote. Offshore oil leasing is continuing along the Atlantic coast, despite misgivings about ocean pollution from tanker accidents, oil spills, and oil well blowouts like the 1979 accident off the coast of Mexico. Yet Congress resisted the formation of an Energy Mobilization Board proposed by President Carter to avoid delays in approval of major energy projects. The Reagan administration is more willing to accept environmental damage for the sake of energy expansion and economic growth. Environmental regulations have been weakened, mainly by drastic cuts in EPA budgets and by reluctance to enforce standards, rather than by legislative changes. Congress and the public seem willing to accept some postponement of environmental quality goals but are likely to resist any major lowering of air and water standards.

Nuclear and Solar Impacts

The nuclear fuel cycle has by far the smallest ecological impact of any current energy option. Uranium mining leaves some tailings, but in much smaller quantities than coal. With breeders, the volume of tailings would be even smaller, and the land area disturbed by mining would be negligible. Nuclear plants produce no air pollution. While their thermal pollution is higher than that from coal (since no heat goes up the smokestack), the heat can be dispersed by cooling towers rather than by discharge into rivers, lakes, or oceans. The use of land for plants and transmission lines is comparable to that with coal.

The environmental impacts of solar technologies vary widely. In most cases they are intermediate between coal and nuclear sources. Large dams create major problems in land use, habitat destruction, siltation, and in some cases eutrophication (algae growth) and damage to downstream aquatic life and soil fertility. For small dams, these effects appear to be smaller relative to their electric output. Solar electricity from mirrors and "power towers" would require large areas, perhaps five or ten square miles for a 1,000 megawatt plant. Photovoltaic panels could make use of rooftops and southern walls, but large blocks of electricity for cities or industries would need large arrays. Only in desert or rocky areas or very marginal land could such arrays be located without damage to vegetation.[29]

Biomass sources can provide fuels to replace oil and gas. Agricultural and forest residues and mill and municipal wastes can be burned or digested to form methane gas (and, in some cases, fertilizer that can be returned to the soil), with relatively little environmental damage. But the use of corn, wheat, or special energy crops to produce alcohol for fuel would take land that is sorely needed for food. Plantations of fast-growing trees have been proposed, and there are some areas where they would reduce soil erosion. In much of the Third World, reforestation is needed simply to provide the firewood that is in such short supply. Aquatic plants for ocean farming might be ecologically sound. Biomass combustion does not release CO_2 beyond that which was absorbed during plant growth, but it can produce other air pollution problems such as wood smoke (though sulfur emissions are low).[30]

There is also environmental damage and energy use in the extraction of materials for the production of solar equipment. For example, photovoltaic panels use steel (which might require coal and iron ore) and silicon (which is a pollutant if dispersed). Some authors have claimed that health and environmental risks from producing solar equipment would be comparable to risks from coal-produced electricity, but others claim that the risks would be smaller than those from either coal or nuclear energy.[31] There is of course no way that solar systems can be used for weapons, and no opportunity for catastrophic accidents except for large dams. The very diversity of solar technologies and their geographical dispersal would diversify the risks. In terms of environmental impacts, solar sources appear in general to be potentially worse than nuclear energy, but probably better than coal, for equivalent energy production.

Even conservation involves risks to health and the environment, though they are relatively small. Conservation to reduce demand is almost always cheaper, and is often safer, than an equivalent increase in supply. But there are environmental and health costs for the materials and energy needed to fabricate and install insulating materials. Some insulation gives off formaldehyde, which causes respiratory illness. Sealing a house tightly against air leakage allows a buildup of radon gas (a cause of lung cancer) from cement, bricks, or soil; this can be avoided with air ventilation and a heat exchanger, but at considerable cost.[32] Smaller cars save energy and pollute less, but the fatality rate in accidents is higher. Most forms of conservation entail environmental and health risks as well as economic costs, though in most cases these are less than for equivalent energy production.

WORLD PEACE

Of all the risks from energy choices, nuclear war would clearly have the most disastrous consequences. Local conflicts between two smaller nations might escalate to involve other nations. Once small nuclear weapons were used by major nations, even for "tactical" or "limited" purposes, it is unlikely that the side facing defeat could resist using more powerful weapons. The destruction and radioactivity from an all-out nuclear exchange would be so widespread that any "victory" would be empty. Such a war in the name of freedom would be ironic indeed, since widespread catastrophe would be likely to engender more authoritarian governments. A nuclear holocaust might conceivably threaten the survival of humanity itself.

According to proponents of the "just war" theory, a war is just if the cause for which it is fought is just and if attack is limited to primarily military targets. In a just war there may be noncombatant victims, but they must not be the intended target and their numbers must not be "disproportionate." A counter-city nuclear attack with at least 20 million casualties—plus indirect and delayed effects—could not conceivably fulfill these conditions. Even a counter-force strategy, aimed at military installations rather than population centers, would have a very high civilian casualty rate, in violation of the proportionality rule.[33]

Nuclear war would be devastatingly destructive to the United States as well as to other countries. For the past 30 years we have relied primarily on "mutual deterrence" to prevent the use of nuclear weapons by the superpowers. For deterrence to be credible, we must intend to retaliate if we are attacked; yet after an attack, retaliation would serve no purpose except to destroy additional millions of people. Malfunction of equipment or miscalculation of an opponent's responses might also trigger an unexpected and uncontrollable sequence of events. Moreover, the reliability of deterrence is increasingly uncertain as more and more nations acquire nuclear weapons capability. Terrorist groups have used bombs or bomb threats with growing frequency, and retaliation would be ineffective against nuclear threats by such groups.

There are, of course, many actions unrelated to energy policy that would reduce the probability of nuclear war. Arms control agreements, such as a comprehensive test ban and the extension of the SALT agreement with adequate verification, could slow and then reverse the nuclear arms race, which provides uncertain security at such enormous cost. Resentment over the growing gap between rich and poor nations is a continuing threat to world peace and stability and a fertile soil for the spread of Marxism and Russian influence. Efforts in international assistance and cooperation and the development of a more just international economic order might reduce the incentives to use nuclear weapons. There are two aspects of energy policy, however, to which the prospect of nuclear war is relevant: nuclear proliferation and international conflicts over oil. Both are likely to be crucial issues during the 1980s.

Nuclear Weapons Proliferation

In the past, nations have manufactured nuclear weapons prior to nuclear power programs or independent of them. Weapons can be made from current reactor fuels by uranium enrichment or the extraction of plutonium from spent fuels, but these are high-technology operations that are difficult to carry out secretly. Widespread deployment of reprocessing or breeders around the world would make weapons production easier because large quantities of separated plutonium would be in circulation. It can be handled with relative ease, and theft during storage or shipment would be tempting to small nations or revolutionary groups—whose actions would be unpredictable. For such reasons, several recent reports regard diversion and proliferation as the most significant issues in nuclear energy policy.[34] The Carter administration delayed breeder and reprocessing deployment indefinitely because of such considerations. The Reagan administration favors the deployment of both breeder and reprocessing technologies. It has also advocated the reprocessing of U.S. commercial spent fuel to obtain plutonium for nuclear weapons—which is precisely what we have tried to prevent other nations from doing.

Nonproliferation negotiations to date have had very limited success. Several nations have never signed the Nonproliferation Treaty, and the nuclear weapons states have failed to negotiate nuclear arms reduction, as the treaty requires. France, Great Britain, and the Soviet Union are now going ahead with breeders and reprocessing. The International Nuclear Fuel Cycle Evaluation supported breeder development, though its estimates of energy demand and uranium supply through the end of the century suggest that there is less urgency than most of its participants had assumed. The study holds that security measures and international inspection safeguards could be strengthened somewhat, and accounting procedures for plutonium inventories could be improved, but the diversion of small amounts would still be likely to escape detection. While the colocation of breeders and reprocessing plants in guarded "energy parks" would avoid the shipment of plutonium, surreptitious diversion by national governments themselves would be difficult to prevent.[35]

One alternative would be the development of proliferation-resistant fuel cycles that are more efficient than current reactors but less susceptible to proliferation than breeders. With partial separation, some uranium would be left after reprocessing so that the fuel is not directly usable for weapons, or some fission products would be left so that it is highly radioactive. Advanced converter reactors are particularly promising since they are efficient enough, even without reprocessing, to extend the use of high-grade uranium deposits by a century. They could later be used even more efficiently with reprocessing, running on denatured uranium, a mixture of isotopes that is not of weapons grade. Decisions about reprocessing might thus be postponed for several decades.[36]

Another possibility is the "symbiotic cycle" in which denatured uranium-233 would be shipped from breeders and reprocessing plants in secured areas to dispersed converter reactors; the spent fuel would be returned for reprocessing. Ideally the secured areas would be regional centers under international management. If coupled with a vigorous effort at reduction of nuclear arsenals, this would offer a new opportunity for international control of nuclear weapons, but it faces formidable political obstacles. The provision of waste disposal services would be an added incentive for some nations to cooperate. Even without international management, the confinement of sensitive facilities to a few regional centers would make international inspection easier.[37]

In sum, the potential contribution of nuclear plants to energy supply goes up and proliferation-resistance goes down in the following sequence of options: (1) closing all nuclear plants, (2) continued use of light-water reactors, (3) advanced converters, and (4) breeders. In all cases, the extent of proliferation would depend on actions by other countries as well as United States policies. While our example and influence may have some affect on decisions by other countries, any major reduction in proliferation risks requires international agreement and inspection. Since clandestine enrichment or reprocessing facilities could provide alternative routes to nuclear weapons, our decision on breeders will probably only slightly alter the overall prospects of weapons proliferation. But it can be argued that even a small reduction in the probability of nuclear war should be sought, if it can be achieved without energy shortages that would create other kinds of international conflict that would be even greater threats to world peace.

Energy Shortages and Oil Wars

The United States, Western Europe, and Japan are heavily dependent on imported oil and highly vulnerable to the interruption of supply. President Carter declared that access to Middle Eastern oil is so vital to our national interests that we would use military means if necessary to keep it flowing, and President Reagan is even more likely to take such action. The Soviet Union is deeply involved in the struggle for oil and in the politics of the Middle East. Severe energy shortages in other nations could lead to economic and political destabilization, unemployment, violence, and revolution; desperate nations might resort to nuclear threats as a form of blackmail.

The Middle East is an area of complex rivalries, political instability, and rapid social change. We have supplied arms to Israel, and also to several Arab States (to retain their goodwill, to counter Soviet influence, and to reduce our balance-of-payment deficits from oil). Our largest oil imports come from Saudi Arabia, whose future is very uncertain. Will the royal family retain control in the midst of neighboring revolutionary movements and the domestic conflicts created by the sudden modernization of a traditional society? Will they reduce oil production if oil in the ground seems a more valuable asset than bank accounts in industrial nations with high inflation rates? Will they continue to be a moderating influence in OPEC, or will they press for ever higher oil prices, which are creating havoc in the world economy?

The reduction of United States dependence on foreign oil would thus be a major contribution to national security and to world peace and security. Domestic oil production can be raised somewhat, but the increase is not likely to be large. Despite increased drilling stimulated by higher prices, proven U.S. reserves have continued to decline, and there is little prospect of finding major new domestic fields. Oil shales can also be used, but at high economic and environmental costs. Autos and trucks account for 75 percent of our oil use. Increased auto efficiency, effective measures for the conservation of transportation fuels, and the development of synthetic liquid fuels should have high priority on the national agenda. Oil is used for the generation of 12 percent of our electricity (mainly in New England, Florida, and California); these plants could be converted to coal or replaced by coal or nuclear plants. The enlargement of the strategic petroleum reserve in salt domes in Louisiana and elsewhere would reduce our short-term vulnerability to supply interruptions. With a standby gasoline rationing scheme we could survive a cutoff for a few months, but economic disruption would be immense thereafter.[38]

In the short term, the dangers of war from the escalation of conflicts over oil appear greater than those from nuclear proliferation. In the intermediate term, any judgment depends on estimates of other energy sources. Coal use can be expanded considerably, but it entails major environmental and health hazards, and it is very unevenly distributed among nations; the United States and the Soviet Union have 74 percent of known world reserves, while all of Africa and Latin America together have less than 1 percent. Solar technologies could play a limited role in the intermediate run and a larger role in the long run, if costs can be brought down. The materials needed for diverse solar technologies vary greatly in abundance, but we noted earlier that, unlike oil and coal, sunlight is rather evenly distributed. Solar energy and perhaps fusion may eventually help to reduce international competition for fuels. In the meantime, proliferation resistance and reduction in oil dependence must be taken with the utmost seriousness along with efforts to halt the nuclear arms race. World peace is the most important but also the most difficult issue we face during the 1980s.

LONG-TERM SUSTAINABILITY

Our energy policies affect future generations. In a few decades we will have exhausted the oil and gas that took hundreds of millions of years to accumulate. Our radioactive wastes will be dangerous for periods as long as 100,000 years unless they can be effectively isolated from human populations. In such cases, what obligations do we have to future generations?

The economic and the political systems both have very limited time horizons. Industry is interested in rapid return on investments. Political leaders have difficulty looking beyond the next election. Economists apply time discounts to future costs and benefits, effectively ignoring consequences more distant than two or three decades. Depletion of irreplaceable resources transfers additional burdens to our descendants, who cast no votes today. But there are several ways in which our time horizon might be extended.

In previous centuries, most Americans saw themselves as building for the future, creating a nation, and working so that their children could have a better life. Early in this century the conservation movement was dedicated to careful maintenance of soil, forests, and other natural resources as a national heritage for the future.[39] The National Environmental Policy Act of 1969 declares that each generation is "trustee of the environment for succeeding generations." At a time when we are preoccupied by immediate crises, which we think are susceptible to quick solutions, we need to recover the idea of working for a long-term national future, seen now in a global context.

Our religious traditions have also asserted that we have obligations to posterity. Both Judaism and Christianity have expressed a universalistic vision of the unity of humankind embracing generations yet to come. In the early biblical view, stewardship requires consideration of posterity because God's purposes include the future. The land, for example, was to be held as a trust for future generations. This long time-horizon derives from a sense of history and an orientation toward the future, as well as accountability to a God who spans the generations. To be sure, the time perspective changed in the early Christian period when the imminent end of the world was expected. Later it was held that the Kingdom of God could be partially realized on earth, though fully realized only by God's action or in heaven. But recent theologians have reasserted the theme of obligations to future generations. Both the National Council of Churches and the World Council of Churches have defended sustainability along with justice and participation as criteria for energy policy.[40]

Another source of long time-perspectives is ecology. Ecologists commonly study population changes over many generations. They are aware of the interdependence of the components of systems and the indirect consequences of actions distant in time and space. The ecologist is interested in the long-term carrying capacity of an environment, the sustainability of its resources, and the survival of its life forms. The study of evolutionary history is also conducive to a wide temporal horizon. Much of the recent concern for future generations can be attributed to the influence of the environmental movement. In this framework it is evident that population growth is a major threat to the world's future, and population stabilization should have high priority. Population size will be an important factor in future energy demand.[41]

Finally, philosophical principles provide more specific guidance for decisions affecting future generations. The principle of utilitarianism, "the greatest good for the greatest number," is usually taken to include future persons. According to utilitarianism, one should choose the option that maximizes the total good, integrated over time and space. However, the principle is difficult to apply in practice when the distant consequences of our actions are so uncertain. Moreover, consideration of an indefinitely long future might completely dominate over more immediate consequences, unless a time discount is applied, which does not always appear justifiable.[42] In addition, utilitarianism faces difficulties when population size is itself a policy issue. In principle, *total* welfare might be maximized by having a very large population at a low level of welfare. Some utilitarians have proposed an alternative criterion, the highest *average* welfare; but this goal might be achieved by allowing only a very small population with a high level of welfare.[43]

A more plausible approach to intergenerational equity can be found in an extension of the ideas of John Rawls. In the previous chapter, we noted his suggestion that principles of justice can be derived by imagining a social contract between persons who do not know in what social position or generation they will live. Rawls assumes only contractors now living. If instead they are imagined to be drawn from all generations, they will have to decide what policies they would favor in the distribution of resource use over time. All generations would count equally, with no discount for remoteness in time.[44] In order to know which is the least advantaged generation, we would have to balance future resource depletion and environmental degradation against the advances in technology with respect to which future generations will be better off.

In practice it is impossible to consider an indefinitely long series of future generations in policy decisions. But in the case of *renewable* resources, the same result can be obtained by aiming for the *maximum sustainable yield.* Once the sustainable yield is exceeded by excessive fishing, cutting, over-grazing, or soil erosion, for example, the productivity of oceans, forests, grasslands, and croplands is rapidly reduced. People are then consuming productive biological "capital," rather than living on the "interest" that could continue indefinitely. A fair distribution over the generations (assuming population stabilization) can be achieved by keeping within sustainable yields, which can be calculated from knowledge of ecosystems today. Future sustainable yields may be somewhat higher (due to technological advances) or lower (due to environmental damage); but use levels could be readjusted in the light of new information.[45]

With respect to *nonrenewable* resources, equity between generations would require that *the resource base* should not be depleted more rapidly than it can be extended by technology, since all generations have an equal claim on it. Technology turns useless raw materials into useful resources. Those who deplete resources should compensate future generations by passing on improved technology and capital investment to offset the effects of depletion. If the resource base relative to technology is thus preserved from one generation to the next, it will be preserved for all generations. Brian Barry suggests that the criterion should be equality of opportunity with respect to productive potential. The present generation should pass on to the next the technological improvements that would replace the productive opportunities lost by resource depletion.[46]

In the past, technology has extended the resource base at a rate that has more than offset depletion. New extraction technologies have made lower-grade ores economical, discoveries have expanded reserve estimates, and substitutes have been found for scarce materials. As a result, the price of most raw materials (in constant dollars) stayed constant or fell from 1900 to 1970. But during the 1970s, shortages and rising prices affected several materials as the more accessible reserves were used up. We have used oil more rapidly than technology has extended oil reserves, though we have also been developing other energy technologies that might replace oil. Technological knowledge is itself a significant legacy to posterity, but the indirect costs of technology in the form of toxic wastes, pollution, and environmental degradation pass on an increasing burden of risks. The effects of a toxic chemical may not even be evident for several decades. Yet improvements in industrial processes and in pollution control technology could in many cases offset these threats to health and the environment.

Risks passed on to future generations are of special ethical concern for two reasons. First, they are imposed involuntarily, since future generations cannot give voluntary consent or have any voice in current decisions. Second, there is a distributive inequity if risks to future generations arise from actions of which the present generation is the main beneficiary (as in the case of radioactive wastes). To be sure, resource use contributes to technological development and economic growth. But we must remember that 90 percent of natural resources go into short-lived consumer goods, and only 10 percent into capital accumulation and technological equipment from which future generations might benefit.[47] The costs of reducing future risks should thus be borne by the present generation.

Concern for the needs of future generations must of course be combined with concern for the needs of those now living. Urgent current needs have priority over uncertain future ones. But the basic needs of future generations should have priority over the luxuries of the present. Brian Barry points out that the transfer of technology and capital from industrial nations to LDCs would fulfill intergenerational and international justice simultaneously. For our obligation is not simply to our own descendants, but to future generations everywhere—and particularly to those who missed the chance to industrialize while energy and other resources were cheap. Barry proposes an international income tax, which would fall most heavily on those with highest income, combined with a severance tax on the use of nonrenewable resources, which would fall primarily on those who consume the largest quantities of resources.[48] Such taxes may be difficult to effect, but they show that justice across space and justice across time are not incompatible.

Our present political and economic institutions virtually ignore any considerations beyond the present generation. We have been shortsighted in giving depletion allowances and other subsidies to encourage rapid extraction of oil and other nonrenewable minerals, whereas until recently we gave little support to sources that are renewable or sustainable for a long time. What policies would appear reasonable if we took both present and future needs into account?

Sustainable Sources

Coal reserves would last for several centuries at current rates of use. Even at accelerated rates they might last for more than a century, but the CO_2 buildup would probably be critical long before that, and other environmental impacts would be serious. We can count on coal only for a few decades during the transition to sustainable fuels.

Nuclear fusion is still at the experimental stage, but it holds out the prospect of long-term sustainability, since its main fuel is heavy hydrogen, which can be obtained from ordinary water; the techniques that are currently most promising also make use of a lithium isotope that is more difficult to obtain. Fusion would produce no significant environment damage or air pollution, relatively little radioactivity, and no weapons-grade material. Yet there are formidable scientific and engineering obstacles in maintaining temperatures of millions of degrees, and economic costs are very uncertain. Research on fusion should be vigorously pursued, but it is too early to know whether it will work in the laboratory, much less whether it will be practical on a commercial scale.[49]

This leaves two current technologies that are sustainable: breeder reactors and solar energy. Breeder reactors are highly efficient, extracting 60 times as much energy from uranium as light-water reactors do, and extending reserves by making low-grade ores economical. Uranium and thorium reserves are not strictly inexhaustible, but with the breeder they would last for many centuries—or longer if even more dilute sources such as seawater can be used. But we have suggested that proliferation risks are serious, and that until nuclear weapons are brought under control, advanced converters represent a possible compromise between efficiency and proliferation resistance. The technologies of breeders and reprocessing have been demonstrated, but there is considerable uncertainty about their economic costs.

Energy from the sun is inexhaustible on the scale of human history. Solar energy used today does not lessen that available to future generations, though solar equipment uses exhaustible materials, and renewable biomass sources deplete soil nutrients unless cautiously used. Solar hot water and space heating is already competitive with oil or gas in many situations, figured over the life of the system.[50] There are no fuel costs, but initial equipment costs are usually too high to be within the reach of low-income families unless helped by low-interest loans repaid from fuel savings. The National Center for Appropriate Technology has been developing low-cost solar methods suitable for urban and rural self-help projects.

The extra cost of solar design in new homes will pay for itself more rapidly as fuel prices rise. Passive heating requires glass-covered south walls and thorough insulation. Active heating involves solar collectors and the circulation of air, water, or alcohol. The main barriers currently are institutional rather than technical. Building codes are only slowly adapting to solar energy, and city ordinances are only beginning to provide protection against nearby construction that would block access to the sun. Many utilities will not grant "all-electric" rates for electric heating as a back-up for solar systems, even though no other fuel is used. They have also tried to discourage on-site electricity generation and until recently have refused to buy surplus electricity from local sources, which could feed back into the network. Loan companies have been slow to acknowledge the resale value of solar homes. Most of these obstacles could be reduced by appropriate legislation at local, state, or national levels.[51]

Biomass is another important energy source. Current wastes are the best place to start; forest and agricultural residues and urban trash can be burned, fermented to make alcohol, or digested to make methane. Next, wood can be burned in residences and industrial boilers; with careful management and reforestation, increased yields on half of the commercial timber land could be used for fuel. New techniques for making wood pellets offer the prospect of much lower transportation costs. Gasohol from corn or wheat appears less promising and would tend to raise food prices. Small dams, windmills, and geothermal sources are also practical in some parts of the country.[52] None of these sources is very large, but together they could make a sizeable contribution. Estimates of total U.S. solar potential vary considerably. The Council on Environmental Quality claims that at least 50 percent of the nation's energy could be solar by 2020. Recent estimates for the year 2000 range from 8 percent to 35 percent solar; an intergovernmental task force set a goal of 20 percent. The range of estimates reflects differing assumptions about technological advances, prices of competing fuels, and government policies in support of solar development.[53]

The two most significant advances would be improvements in the efficiency of photovoltaic panels and of plant photosynthesis. Both would lead to simultaneous reductions in cost, land use, and material use. The cost of photovoltaics has fallen somewhat and is expected to fall rapidly as mass production processes are introduced. Better storage systems are also needed since both sunlight and wind are intermittent. Promising lines of research include pumped storage reservoirs, advanced batteries, and hydrogen electrolysis. In the more distant future, recombinant DNA techniques may lead to organisms or plants that make more efficient use of sunlight.

Even when current solar costs do not appear competitive to the consumer, solar energy may be a wise investment for society. First, the cost of most fuels is rising, while solar costs can be expected to fall with further research, mass production, and new government policies. In the case of solar electricity, costs to society should be compared not with the *average* cost of electricity from existing plants, but with the *marginal* or replacement cost of electricity from new plants, which is considerably higher. Electricity rates are currently regulated at far below the replacement cost.

Second, market prices still omit many environmental and human costs borne by society, even when some of the externalities have been internalized through environmental and safety regulations. We have seen that coal use, in particular, has substantial impacts on land, health, and perhaps climate. The market takes a short-term view and discounts costs to future generations. As accessible nonrenewable resources are used, benefits are transferred from the future to the present, while some risks are transferred from present to future.

Third, other fuels have been heavily subsidized in the past—adding up to at least $120 billion in the last 20 years.[54] Oil has been subsidized by depletion allowances, tax credits, import quotas, and highway construction funds. Regulated prices gave oil and gas a competitive advantage and encouraged the technologies that use these fuels. Various nuclear subsidies have been mentioned: past R & D funds, present insurance coverage, and perhaps some of the future costs of accidents (such as TMI), waste disposal, and the decommissioning of worn-out plants, which may not be adequately covered by the funds set aside by utilities for these purposes. Because the energy industry has huge investments in fossil fuel and nuclear technologies, most of their research—and their political influence—supports continued use of these fuels. Less than 1 percent of U.S. energy R & D since World War II has gone to solar research.

There are good reasons, then, for funding solar research and for giving solar tax credits. California gives a 55 percent tax credit on expenditures for solar equipment, and in 1980 Congress approved a 40 percent solar tax credit.

The Harvard Business School study advocates 60 percent tax credits for solar purchases for a few years (gradually phasing out), to represent their real value to society, including reduced dependence on imported oil. The study defends reliance on the free market but argues that it is necessary to correct for market distortions introduced by previous subsidies, regulations, and international politics.[55] We could also say that since solar energy is a truly sustainable source its development would benefit future as well as present generations. If we did not know in which generation we would live, we would probably give strong support to solar research and development.

Radioactive Wastes and Future Generations

There are two kinds of radioactive wastes from nuclear energy, both of which involve very low levels of risk lasting for very long periods of time. First, uranium mine and mill tailings are rich in radium, which can contaminate groundwater, and they release radon gas, which can cause lung cancer. In the past, tailings have been piled in abandoned sites where they are susceptible to water and wind erosion. The isotope that produces radon has a half-life of 80,000 years, so that fatalities, even at low annual rates, would mount up over the years. There is an inequity in exposing thousands of future generations to risks for the sake of benefits to the present generation. In this case, covering the tailings with a layer of earth (or clay, which reduces the chance of reexposure) is an effective seal against the escape of radon. Subsurface burial or backfill into mining pits or deep lakes would provide even more secure long-term protection.[56]

Second, nuclear plant wastes are much smaller in volume but far more intensely radioactive than tailings. With no reprocessing plants operating, spent fuel rods have been accumulating in pools at power plants, whose storage capacity will soon be strained. Most of the radioactivity in spent fuel comes initially from fission products, but that remaining after a few centuries comes mainly from transuranic elements, especially Plutonium with a half-life of 24,000 years. Only after 250,000 years would the activity of such wastes be comparable to that of natural uranium ore. Reprocessing would remove most of these long-lived elements for recycling as reactor fuel. The liquid wastes from reprocessing are more difficult to handle than fuel rods, but the very long-term repository risks are lower. When solidified, their levels of radioactivity after 1,000 years would be comparable to those of uranium ores.[57]

Retrievable storage is probably desirable while decisions about reprocessing are still pending, and the first deep underground storage should perhaps be retrievable so it can be monitored and modified if necessary. But in the long run, would future generations favor retrievable or irretrievable disposal? Only a small fraction of the energy potential of nuclear fuel is currently used; even after reprocessing, additional energy and materials could perhaps be extracted from wastes by technologies not yet discovered. But accessible storage entails greater long-term risks and burdens of management. Alvin Weinberg has advocated retrievable underground storage with tight security measures to guard against flooding, drilling, theft, or sabotage. He describes the "Faustian bargain" of energy growth at the price of "a vigilance and a longevity of our social institutions that we are unaccustomed to." He favors the bargain and says that it would require a "nuclear priesthood" of technicians with high discipline to maintain essentially perpetual surveillance over buried wastes.[58] But is it realistic to count on the stability of social institutions on a time scale of 1,000 years, much less 250,000 years? No social order in history has lasted more than a few centuries.

One of the criteria proposed by EPA for the isolation of radioactive wastes is that "controls which are based on institutional functions should not be relied upon for longer than 100 years to provide such isolation."[59] But can we guarantee the stability of social institutions for even a century in the United States? What about other countries that are likely to adopt the technologies developed in the United States? The occurrence of two world wars and many smaller wars and revolutionary upheavals already in this century suggests that in a fast-changing world we should use methods that do not depend on human monitoring and protection over periods of several generations.

If wastes are solidified and deeply buried, the chief risk is leaching by groundwater into aquifers and thence into drinking water and food chains. Contrary to public fears, there would be no possibility of explosion or sudden release, and sabotage or diversion would be virtually impossible. But groundwater patterns might be altered by human intervention, climate changes, seismic activity, or undetected fissures. The leach rates from wastes embedded in ceramics appear to be very low, though detailed research on reactions among particular combinations of elements is only just beginning. Some reassurance is provided by the slow migration of the fission products from the natural chain reaction, which occurred several million years ago in uranium ores in Okla, Gabon, though the geological conditions and reacting elements in a disposal site might be rather different.[60]

The problem of permanent disposal received little attention in the 1950s and 1960s. In the early 1970s, plans called for burial in salt mines, but one site was abandaned when old drill holes were found, and at another site pockets of brine were discovered. It now appears that granite or basalt may offer better protection than salt beds. Yet there still are many unanswered questions about geological stability and the interaction of wastes with water and rocks at high pressures and temperatures. A 1978 National Academy of Sciences study calls for more extensive research on alternative solidification methods and burial sites. The study holds that there are promising options, but that we do not know enough yet to make such an important decision, and safe disposal is likely to be costly. An interagency review in 1979 concluded that there are still too many uncertainties to start building a permanent facility.[61] Criteria for site selection and standards for hazards from radioactive wastes are still being developed. The Reagan administration favors expansion of nuclear power but has not yet formulated a policy for waste disposal.

The most immediate problem is the need to restore public confidence in federal management of radioactive wastes. For 25 years, wastes were treated as a minor technical question that could be deferred; political, social, and ethical issues were ignored. There was little opportunity for public participation in the formulation of disposal policies. Public confidence can be restored only by a broad program of research and site studies with a diversity of solid forms and geological media. There should be wide opportunity for public participation, access to information, and independent scientific review. It is also important for state governments to have a role in site selection, though a state should probably not have veto power over the location of a repository within its borders.[62] As indicated earlier, regional equity suggests that a state should receive compensation for the costs and risks it would assume for the sake of beneficiaries elsewhere. Present beneficiaries should pay the costs of safe disposal, since there is no way in which they can compensate future generations, apart from a general legacy of technological progress, which should not become an excuse for justifying preventable risks of extremely long duration.

To sum up, each of the current energy options entails both short– and long–term risks. In the case of oil there is vulnerability to price manipulation and embargo and the risk that international conflict could escalate to war. In the long run, of course, oil reserves will be exhausted. With coal, the risks from SO_2 and acid rain can be greatly reduced by installing stack scrubbers, but at substantial cost. Even with strict mine safety and land reclamation standards, coal mining exacts a human and environmental toll. After a few decades, the buildup of CO_2 may produce major climate changes. With nuclear power, the probability of a reactor meltdown is very small, but the consequences would be very large. The spread of breeder reactors without effective international safeguards would make the diversion of material for nuclear weapons somewhat easier. Radioactive wastes appear to be a manageable problem, but we have yet to devise a system whose safety can be assured over long periods of time. With solar energy the main short-run obstacles are the high costs, especially for the production of electricity, and the shortage of land for biomass to produce fuels. But the long-term prospects for a variety of solar technologies appear promising.

Two policy recommendations follow from the conclusion that each of these sources is associated with significant though often uncertain risks. First, a strong program of conservation is crucial in reducing the growth of energy demand in order to minimize the expansion of energy supply. Second, a mix of supply technologies is desirable to diversify the risks, to keep our options open while some of the technological, economic, and environmental uncertainties are resolved, and to facilitate the transition to sustainable sources. These recommendations are discussed in the final chapter of this volume. Before considering them, we must examine further the tradeoffs among the values presented in this chapter and the previous one and look at the dilemmas of implementing policy choices in practice. We must also analyze the context of national and international politics within which policy decisions will have to be carried out.

6 Values and Policies: Sociopolitical Dilemmas

INTRODUCTION

Few people would quarrel with the statement that the choice of energy technologies should reflect the ethical principles accepted by a society and the political values of that society. The problem comes in agreeing on the implications of specific choices for specific values and on the relative importance of different values when they cannot be simultaneously maximized by any technological choice. Yet this is not the problem that is usually addressed. Rather, the advocates of one technology choice accuse the advocates of another of being "unethical" and insensitive to shared political values.

The values involved exist along many different dimensions. They include:

- Social equity versus economic efficiency. The protection of the environment versus economic growth and material welfare.
- The conservation of valuable resources for future generations versus creating the capacity to solve future problems.
- The imposition of risks or costs on people without their consent versus the paralysis of decision making through competition between divergent interests.
- Delegation of major social decisions to bureaucracies or technical experts versus participatory democracy.
- The responsibility of the rich nations for the poor nations and how that responsibility should be exercised in relation to the sharing of natural resources.
- Material consumption versus frugality and their moral effects.
- Sense of community versus individualism.
- The "public interest" versus individual welfare and preferences.

- The relative social and political merits of simple versus complex technologies.
- The appropriate relative emphasis on present welfare versus that of future generations.

It is not only that people place different weights on these social values and tradeoffs, but the implications of concrete policy choices along each of these dimensions of value choice are by no means self-evident.

Thus the advocates of both "hard" and "soft" energy technologies accuse each other of wishing to impose authoritarian values on society through their choices. The advocacy of "hard" energy technologies, whether of nuclear power, large central coal generating plants, or transmission of electrical energy from large solar installations in the Southwest desert is identified with authoritarian and conservative politics, while antinuclear activism and the advocacy of energy conservation and localized renewable energy sources is identified with left totalitarianism and the desire to impose elitist preferences on the whole society while ignoring the plight of the underprivileged. Organizations representing workers and blacks have come out in favor of nuclear power as the only route to hope for the underprivileged,[1] while other worker and minority organizations have looked upon nuclear energy as a conspiracy to exploit the poor and sacrifice their health and safety to the enrichment of the capitalist few.[2]

The actual impact of energy policy on political values, however, is not a matter of intention alone, or something that is self-evident from the characteristics of the technology and easily judged without searching analysis. On the contrary, the relation between policy choices and value choices is perhaps the most subtle and difficult issue of all, in part because it involves predictions about human behavior and human responses to concrete legislative or executive actions that are not amenable to rigorous analysis. The realities of political and bureaucratic implementation of policy may produce consequences different from or even opposite to the original intent. Even judgments regarding the societal risks associated with a given technology are intimately dependent upon assumptions about human behavior and the future political and social climate.

This last point is nowhere better illustrated than in the case of nuclear energy. Indeed, the most technically sophisticated critics of nuclear power are prepared to concede that there are no insoluble technical problems associated with its safe deployment. They argue, rather, that human institutions and individuals are simply not up to the task of safely managing a large-scale nuclear industry, and that the "worst-case" risks of mismanagement are so high that no imaginable benefits are worth incurring such risks, however improbable they may appear.[3] The proponents of nuclear power, on the other hand, point to the rapidly improving safety record of international air transportation and argue that this industry involves far more people in a less disciplined organizational setting, and a technology in many ways even more complex and unforgiving than nuclear technology.[4]

On the other side, the critics of the renewable technologies argue that while health and safety considerations are not as obvious or as dramatic as in the case of nuclear power, they may actually be much more pervasive if the implications of the technology are traced back into the supporting industries and forward into the servicing industries and the actual implementation of decentralized technologies in the hands of the ultimate consumer.[5] Implicit in such debates is an argument as to just what constitutes part of the "system" to which the renewable technology belongs. For example, should the dramatic increase in fatalities associated with the use of chain saws since the popularity of wood burning be chalked up against this renewable "biomass" technology (i.e., organic material used as source of fuel)? Is splitting wood really safer than splitting atoms? Or are there different values associated with the "voluntary" risk of the householder who kills himself with a chain saw in comparison with the "involuntary" risk of the family that happens to live near a nuclear plant?

It is also true that certain technologies acquire a symbolic political or ethical significance that may be much more important than their "objective" risks or benefits.[6] Moreover, there is a wide disagreement among analysts as to what is really "objective." There is an implication in the use of the word that suggests that people's feelings toward a technology are somehow less legitimate and are not "facts" in the same sense as the increased risk of cancer from exposure to low level radiation. Individuals with a scientific or technical background are especially inclined to dismiss these "subjective" considerations as inadmissable criteria for the choice of energy technologies. They see some justification for this view in the fact that such political valuations change with time, are heavily influenced by media attention and mode of treatment, and are strongly conditioned by the level of awareness of the benefits of the technology and the availability of viable alternatives. Many nuclear critics, for example, think that solar energy is just around the corner in a matter of a few years and is only delayed by a conspiracy among the purveyors of traditional energy supplies, particularly the major oil companies.[7] The fact that this analysis has been shown to be gravely flawed has not vitiated its credibility in many quarters.[8]

Other policy analysts argue that public perceptions about a technology are just as "real" as the physical facts about it, and that in the early stages of the deployment of a new and untried technology "objective" risk assessments can often change as rapidly or more rapidly than public perceptions as new information or research results accumulate. Especially in the past era of euphoria about technology, the proponents of new technologies have tended to overlook evidences of possible risks until the accumulation of evidence became too great to ignore.[9] Scientific critics of nuclear safety have been sufficiently right sufficiently often to raise as much doubt about the stability of the scientific consensus as the technologists raise about the stability of the public opinion consensus.

These analysts lay particular stress on the importance of process in achieving a wide public consensus on the choice of technology. Unless choices can be made through a political process that is widely perceived as legitimate, they say, the political view of the safety or benefit of a technology will override the technical view, and the technical view can only prevail at the cost of subverting democratic institutions. In other words they assert that political side effects of pushing the prevailing technical consensus in the face of contrary public perceptions, however ill-informed technically, are likely to be even more serious than any loss of social benefit due to the nonacceptance of the technology. These analysts would, perhaps, not put things in quite that way. Rather they would tend to rationalize the political consensus by saying that the instincts of the people may after all turn out to be better than the narrow and time-bound professional judgments of the technicians.

In the case of nuclear power there are several symbolic overtones or emotional connotations that may be more important in determining public perceptions than any objective characteristics of the technology.

One is the identification of nuclear power with nuclear bombs arising from the common historical and scientific origins of the two quite different technologies. This identification in the public mind has been reinforced by the fact that the military and civilian applications of nuclear energy have always remained in the same agency, not only in the United States, but generally in all countries that have nuclear power programs. The reality of this identification is suggested by the finding that most of the public thinks, even after TMI, that the main hazard of a reactor is that it can explode like an atomic bomb.[10]

A second symbolic value arises from the identification of nuclear power with big industry and big government at a time in which the legitimacy of these institutions is being questioned from a wide range of locations in the political spectrum.

Another consideration is that ionizing radiation seems uniquely threatening to people in comparison with the "objective" risks associated with it. First, it cannot be sensed and can only be measured by esoteric instruments. One can be exposed to a lethal dose without any immediate tangible evidence of damage. Second, the effects of radiation, especially low levels of radiation, appear only after a long time delay. It is theoretically possible for large numbers of people to accumulate a large dose commitment without knowing it and to have to live out the rest of their lives with an increased threat of cancer. That the same occurs with chemicals, and particularly with smoking, seems somehow less threatening, perhaps because more familiar. Third, radiation is the only environmental agent for which possible genetic effects have been well investigated, and genetic damage that can be transmitted for several generations seems peculiarly frightening. It is not generally appreciated that there is no evidence of genetic damage in humans, even among the victims of Hiroshima and Nagasaki and that the genetic threat is only inferred from model experiments on mice.[11]

Fourth, and perhaps most important, much more is known about ionizing radiation than about any other environmental threats, and therefore it has received much wider public discussion than almost any other environmental hazard. There is plenty of psychological data to indicate that known hazards, no matter how small their probability, are much more feared than ones that are potential but that have never been described in detail.[12] As the disposal of toxic chemical wastes attracts more public attention, and as incidents like Love Canal are publicized, the public may change its *relative* valuation of industrial hazards, but the effect of this is likely to be to enhance the fear of chemical risks rather than reduce the fear of radiation risks.

In the case of solar energy the exact opposite situation prevails in regard to the associated symbolism. The sun is "natural" and ubiquitous. It is familiar and easy to understand, even though the conversion of solar energy to useful form may entail very sophisticated technologies. Solar energy is "free" even though its beneficial use may entail a large capital investment. It can be deployed on a small scale and appears, at least superficially, to be readily amenable to personal control. It does not have to be delivered by large and remote bureaucratic organizations. Its scale is viewed as "human." There are no obvious risks associated with solar energy, and what there are seem to be of a voluntary nature, such as falling off one's roof while repairing solar panels. There is nobody to turn off the juice if you don't pay your bills on time, and solar panels are much harder to repossess if you don't pay your installments than an automobile or a refrigerator. Solar technology appears to be amenable to deployment by small entrepreneurs, although the ultimate structure of a mature industry is by no means clear at the present time. Solar energy also seems especially compatible with a rural lifestyle, which has always been a preferred value in American society. All these aspects give solar energy the appearance of a "benign" and politically attractive technology. In this respect it is almost irresistible from every standpoint except cost, and it is at too early a stage of development for the economics to be widely understood by either the public or the experts. Indeed, the economics can be strongly influenced both by the accounting rules that society establishes for evaluating capital investments and by developments in the pricing of alternative technologies, which are greatly affected by political factors that are hard to foresee. Above all, solar energy has become a symbol of an attractive type of society that reinforces many historic American values derived from our frontier history. These features may be more important than any "hard-headed" economic evaluation, especially in the absence of any significant practical experience with the actual use of the technology by consumers. It is an ideal foil against nuclear energy, a technology that symbolizes many values we don't like—bigness, bureaucracy, complexity, mystification, elitism, and nuclear weaponry.

As the solar-nuclear debate illustrates, the actual value implications of various technological or policy choices are more ambiguous than superficial appearances might suggest. Without trying to reach a final assessment, we shall examine several value or policy tradeoffs that appear in the possible implementation of energy policy, analyzing their implications in detail, taking into account as far as possible the most likely practical details of implementation, which can often be more important than the broader characteristics of the choices. In each case we shall attempt to reach a balanced assessment of the alternatives, recognizing, of course, that it is impossible to completely remove one's own values and political preferences from the assessment, no matter how hard one tries to be impartial.

POLICY CHOICES: THE MARKET VERSUS RATIONING

If a commodity comes into short supply one of two responses is possible. Either one can allow prices to rise, causing supply to rise and demand to fall, until supply and demand return to equilibrium, or else prices can be fixed by political decision and the resulting gap between supply and demand filled through a politically determined allocation of supplies among competing claimants. If the price changes resulting from equilibration of the market are large, questions of equity among consumers and between consumers and producers arise, and political and moral pressures to utilize a rationing mechanism in some form are enhanced. The political problem becomes more acute when the price is largely determined by an external political actor such as OPEC, not subject to either market or domestic political control. The situation is further complicated when one considers the dynamics of the adjustment process. In the case of energy, if the internal supply system is already stretched to its physical limit, then the price elasticity of supply is very low in the short term. Similarly, because most reductions in demand require replacements of energy-consuming plants or equipment, the price elasticity of demand is also low in the short term. Furthermore, demand elasticity may be especially low among lower income groups or for regions of the country where space heating forms a large part of the energy load. Hence, the speed of the change in supply is an important variable in determining the pressures for alternatives to the use of the market. In the actual case of energy in the United States the situation is not quite as described above in that the changes in supply in 1973 and 1979 were caused by the political manipulation of the price of imported oil rather than by a physical shortfall in production capacity, but the effect on the domestic supply-demand balance in the United States is equivalent as long as the equilibrium market price determined by domestic supply and demand would be below the ceiling set by the OPEC price.

If we rely on the market to equilibrate supply and demand domestically a number of consequences follow:

1. Especially in the short run, domestic producers receive large windfall profits resulting from the fact that the selling price of their energy resources suddenly exceeds their production costs by a large margin. This applies in principle not only to oil producers but also to the producers of alternate energy sources as well, to a degree corresponding to the ease of substitution of each alternate source for oil. Thus, for example, the windfall profits to gas producers would be nearly as great as to oil producers, while the profits to coal producers or electricity suppliers would be considerably less because their substitution for oil requires large capital investment and takes longer.

2. Low income consumers, and consumers who cannot switch to less expensive fuels except at large investment costs, have their real purchasing power for other goods and services drastically reduced. The very people and institutions that are least able to make the investments or behavioral changes necessary to reduce their energy costs tend to be the ones most hurt by higher prices. For example, the poor generally own second-hand, fuel-inefficient automobiles, which they use mostly for commuting to work rather than pleasure driving. They usually occupy poorly insulated houses without storm windows, or they live in rental housing where the landlord controls fuel consumption.[13]

3. Because of transportation costs for domestic fuels, in a market situation, there will be wide differences in energy costs paid by those close to supplies of domestic fuel and those who are dependent largely on imported oil. This contrast would exacerbate differences between regions in the cost of energy, and would have disruptive effects on economic competition between different regions. In the United States this was a particularly difficult problem for the Northeast, which was heavily dependent on imported oil for electric power generation in addition to relying primarily on petroleum products for space heating.

4. Different energy companies will differ in their degree of access to domestic fuel supplies because of existing contracts and other historical business relationships with supply sources. In general, small independent distributors of oil products will be at a disadvantage in terms of access to cheaper energy supplies as compared with large organizations, if market forces alone rule.

5. When supplies of energy are restricted suddenly, consumers of all kinds will tend to demand more fuel for inventory purposes both in order to hedge against future shortages and to benefit from resale of their stock at a profit as prices rise further. This hoarding of supplies greatly amplifies any price jump that would be produced by supply-demand equilibration alone in the absence of any change in stocks. This is seen by the public as an additional source of unearned windfall profits at the expense of consumers and is, thus, a source of the belief that shortages are contrived for the purpose of maximizing supplier profits rather than being the result of factors beyond domestic control.

The five above characteristics of an entirely free market, especially in the short term, raise severe problems of equity and justice within a society. The fact that there are many who benefit handsomely, and that these are mainly large organizations, aggravates the sense of injustice to those who suffer, who are mainly the weak and underprivileged. This sense of injustice is increased when redistributional effects are the result of events outside of the control of the people affected. Concern for these injustices leads to the demand that prices be controlled by government. But with controlled prices, more energy is demanded than is available, so that the existing supply must be rationed, and the gap tends to widen with time. As it turned out, price controls on oil and natural gas combined with the elaborate entitlements program that equalized prices among consumers regardless of the source of supply became the prime cause of the large increase in oil imports that occurred following the price jump of 1973. The system necessary to ration supplies among myriads of users can be extremely complex and subtle, and its complexity tends to increase with time as more and more loopholes and imperfections in the system are perceived. This leads us to look at the problems of rationing systems. There are several:

1. Prices that are lower than market prices give misleading signals to both consumers and producers of energy. Consumers continue to demand more than they would otherwise and fail to make investments in more energy efficient equipment, which would be cost-effective if market prices prevailed. Producers fail to invest in alternate technologies and supply sources, which would be profitable if market prices prevailed. These effects would be less serious if the supply shortage were a temporary one. In that case there would be no point in restructuring the economy to adjust to an abnormal situation. This is why, in something like a war emergency, some form of rationing may make sense, although even that is arguable. But if the supply shortfall is long-term, or endemic in the whole system, then price controls postpone necessary economic adjustments and may make the final adjustment that must be made

eventually that much more traumatic and disruptive. The price jump of 1979 following on the Iranian revolution was much more traumatic because of the policies of the U.S. government between 1973 and 1979, which held down domestic energy prices, stimulated U.S. imports, and produced a much tighter market than would have otherwise existed in 1979 if American prices had been permitted to equilibrate with the world price.

2. The system of allocation of oil supplies to oil refiners, known as entitlements, was designed to equalize prices to all consumers by blending the price of cheap domestic oil with expensive foreign oil. The effect of this was a de facto subsidy for oil imports, and predictably this subsidy resulted in a large growth of oil imports and Persian Gulf oil dependence between 1973 and 1979. This situation was further aggravated by especially stringent price controls on natural gas, with preferential allocation to households, so that industrial consumers switched to oil thus further raising oil imports.[14]

3. To be successful any fuel rationing system must be centrally managed. The greater the effort to secure comprehensive equity among consumers the more detailed the management has to be. Any rationing system thus necessitates a growing bureaucracy and more and more complex management techniques. The managerial capabilities of such bureaucracies are not up to the task of anticipating all the secondary and higher order consequences of rationing and allocation. Thus there are more and more efforts to fix one set of side effects by introducing additional regulations, which in turn produce new side effects requiring still more regulation, and so on in an infinite regression. As the whole system accretes complexity, the temptations for corruption and the possibilities for sheer incompetence and mismanagement grow even faster, resulting in an unmanageable system, which, because of its unmanageability, probably creates more inequities and a greater sense of injustice in society than might have resulted from strict reliance on the market. In fact, the creation of a rationing system entails the very kind of big government and burgeoning of impersonal and arbitrary bureaucracies that the opponents of large centralized supply sources deplore as one of the deleterious political consequences of relying on these sources.

If the oil supply and demand system were subjected to the shock of a very large cutoff of imported oil, some form of government intervention would almost certainly become necessary. The market system is designed primarily for gradual adjustment to slow changes in supply and demand, but develops all kinds of pathologies when subject to sudden shocks. Nevertheless, it is by no means certain that rationing and allocation by centralized decisions would be the best mechanism of intervention even in situations of large shock. Various analysts have in fact suggested the use of free market prices in combination with windfall profits taxes and rebates to consumers as a fairer and more efficient system.[15]

The free market and rationing represent extremes in a spectrum of possibilities for energy policy. One end of this spectrum, the completely free market, does not really exist in practice, and therefore we have to look also at the implications of the quasi-free market that we actually have. Two important forms of energy, electricity and natural gas, are delivered by regulated public utilities, and the regulatory process results in market distortions whose consequences are quite different in a period of rising energy prices than in a period of falling prices such as existed worldwide prior to 1970. The most important point is that in a rising price situation the regulation of utilities tends to result in consumer prices for gas and electricity that are less than the replacement cost for these fuels. This is because the earnings of utilities are regulated to be less than a certain percentage return on the historical cost of the investments necessary to make these energy forms available to the ultimate consumer. These historical costs are usually much lower than marginal or replacement costs.

In an extreme case the cost of replacement electricity in the American Northwest may be as much as ten times its present selling price in that region, owing to the fact that so much of this electricity is derived from hydroelectric installations that were constructed during the Great Depression, a period of very low construction costs.[16] What this situation really means in the long run is that the revenues from the sale of electricity are insufficient to finance the construction of new generating capacity either to replace obsolete capacity or to take care of load growth. Thus, as a result of utility rate regulation, both electricity and natural gas tend to be overconsumed at the same time that there is insufficient investment in future supply and distribution. Thus, in an inflationary economy, utility regulation has somewhat the same effect as price controls, described above.

The only reason that this situation has not resulted in electricity shortages is that we still have a great deal of excess capacity derived from a period when the anticipated growth of demand was much higher than at present, owing to low and falling prices. In this particular case the long lead time for energy planning has given us a temporary respite, which may be exhausted beyond the mid-1980s. This situation has led some analysts to suggest that a tax be placed on electricity and natural gas in order to bring the consumer price up to replacement cost. The revenue from this tax might be used both to subsidize consumer investments in more energy-efficient end-use equipment and to subsidize producers for new investments, the optimum division between supply and conservation subsidies being determined by which type of investment has the lowest investment cost per unit of energy produced or saved at the margin. The consumer subsidy might be extended to include investments in renewable energy sources as well as strictly energy-efficient investments, again with choices made on a least-cost basis. The strategy suggested here would thus be somewhat similar to that proposed by the Carter administration for synthetic fuels, in that an excise tax on oil (misnamed a windfall profits tax) is used to accumulate a fund to subsidize synthetic fuel investments and some conservation subsidies. However, in the synfuels case there is apparently no effort to optimize the mix between conservation and supply investments on the basis of least cost, and this has led to political controversy.

There is, of course, a problem with subsidizing conservation investments that is analogous to that we have identified in the case of rationing, although perhaps not as severe. New supply investments represent a relatively small number of large projects, which are fairly easy to monitor and evaluate, whereas energy efficiency investments represent a very large number of small and diverse projects whose evaluation is heavily conditioned by local circumstances. Thus the problem of the cumulative inefficiencies and potential corruption of detailed government interventions arise in acute form. For example, who is to assess the potential energy savings arising from a particular proposed investment, the IRS, which audits the claimed tax credit, or some other agency that directly reimburses the consumer? What are the possibilities for obtaining tax credits or subsidies for improvements that really do not save energy but appear to fall in a category of approved projects? How is one to prevent the wealthier and more sophisticated consumers, including institutional consumers (e.g., large business rather than small business), from benefiting preferentially from such subsidy or tax credit programs because they can pay for advice on how to work the system for all it is worth? Yet such consumer subsidies are inherently more popular politically than supplier subsidies, which go to large organizations. Subsidies to producers are probably more likely to be economically efficient and in principle are easier to monitor, evaluate, and police. On the other hand, it may be easier for large producers to lobby the system to secure preferential advantages; even if they do not actually do so, the suspicion that they are "beating the system" is a serious political liability for any supply-subsidy approach.

The policy tradeoffs sketched above are part of a broader debate in our society about the relative effectiveness and equity of government planning versus reliance on market forces. In this debate it is difficult to sort out the degree to which policy preferences are determined by differences in assumptions about human behavior in response to policies and the degree to which they are determined by differences in political or ethical value preferences. Advocates of market approaches are motivated partly by ideological preferences but partly by empirical evidence concerning the failures of government management of the economy in the past.[17] Usher, for example, has argued that democratic government is inherently unstable in that it will tend to break down if the allocation of income shares becomes too much dependent on political processes rather than quasi-automatic mechanisms such as the market.[18] Thus the load thrown onto political allocation mechanisms by attempting to bypass the market may become a threat to democracy itself. Advocates of a stronger government role are similarly motivated by a mix of ideological and empirical considerations. They tend to be more concerned about the distributional consequences of market imperfections, and a belief that the integrity and coherence of democratic policies cannot survive the inequities and neglect of social costs, which they see as resulting from overreliance on the market. They are also sensitive to the contradictions and inconsistencies in the positions of market advocates. In this situation it is very difficult to tell where values leave off and empirical propositions about human response to policies begin.

EQUITY VERSUS EFFICIENCY

The preceding discussion of the comparative advantages and disadvantages of market vs. rationing mechanisms is actually part of a much broader political issue involving the tradeoff between equity and economic efficiency considerations in social decision making in democratic societies. Economic efficiency is important because it increases the size of the pie to be distributed relative to what would be the case for economically suboptimal decisions. On the other hand, a sense of equity and fairness, or distributive justice, is necessary to secure democratic consent to decisions. Indeed, one of the problems of market mechanisms is that people are less and less willing to accept the social verdict of the market. The impersonality and objectivity of market outcomes is much less taken for granted today in all industrialized societies than was the case 50 years ago, and unwillingness to accept market outcomes without political intervention appears to be growing rapidly. The so-called laws of economics, which were regarded in the same light as laws of nature in the nineteenth and early twentieth centuries, are now seen as human constructs, with ground rules much more subject to manipulation for the benefit of particular interests than used to be believed. The impersonality and neutrality of market mechanisms is no longer taken for granted.

This situation has led economists to suggest a new approach, which attempts to address the questions of equity and efficiency more separately. In fact, economists continually point out that equity issues should never be addressed through price regulation but are more properly addressed through the tax system. Addressing equity through the price system results in large inefficiencies not only because of the huge bureaucratic overheads mentioned in the preceding section, but also because keeping prices low is a blunt instrument, which benefits many besides the people to whom it is targeted. To some extent this latter objection can be overcome by minimal entitlements, examples of which are the food stamp and medicaid programs. In fact, some people have proposed an "energy stamp" program for home heating. "Lifeline" electricity and gas rates for a minimum level of energy consumption per family is another version of such an entitlements approach. All these approaches, however, have a considerable potential for corruption and abuse, if not in fact, then in general public perception, as evidenced by public attention to "welfare chiselers." Clearly the proliferation of such entitlements into more and more goods and services multiplies the potential for corruption and enlarges the bureaucratic overhead. It becomes a divisive element in democracies as more and more interest groups compete for such entitlements. Furthermore, the beneficiaries of controlled prices develop a large vested interest, and entitlements seems to have an inexorable tendency to spread more widely as time passes.[19]

The above dilemma has led to a number of suggestions for procedures that are much more automatic and that provide for a clear separation of distributional and efficiency issues. The classic example is, of course, the negative income tax proposal of Milton Friedman.[20] A more recent example is 1980 presidential candidate John Anderson's proposal for a 50 cent gasoline tax to be used to reduce the employee share of the social security tax.[21] Recently the economist William Nordhaus has made a similar proposal for a tax applicable to all nonrenewable energy forms.[22] Others have proposed energy taxes to be compensated by an increase in the personal exemption or, in the case of people who pay no tax, by the earned income credit. In principle, income redistribution through taxation is completely separable from the price system, and one could achieve the same net distributional outcome without tampering with the market system. Because leaving the price system intact would lead to a more efficient resource allocation, there would be more to redistribute through the tax system and everybody would be better off.[23] What is essential politically, however, is that distributional and substantive policy issues such as energy be handled in a sufficiently coordinated way so that the people affected by both can readily perceive the mutually compensatory effects of the two policies.

This has not been possible in practice in the past. Taxation issues are debated in a totally different forum from energy or environmental issues and attract the attention of different constituencies and interest groups. The recent passage of the windfall profits tax on domestic oil production is probably the first example of an attempt at a coordinated energy and tax policy. Its political life seems likely to be short. For years economists have argued for effluent fees, marketable pollution permits, and other marketlike devices as a substitute for fixed emission or ambient standards and other forms of "command and control" regulation.[24] These arguments have made little headway in the political arena, and less in the United States than in a number of supposedly less market-oriented economies. However, recently there has been increasing interest in marketlike mechanisms for the regulation of pollution, most notably the so-called "bubble concept" recently experimented with by EPA.[25]

The debate between equity and efficiency has a deeper and more fundamental ideological substrate in a debate about growth versus equity. There is one view that says that democracy is only viable in societies in which the economic pie is growing rapidly enough so that political arguments can center on how increments in the GNP are to be distributed rather than how a fixed pie is to be redistributed. The limits-to-growth school, first popularized, though later modified, by the Club of Rome,[26] maintains that the world is already bumping against fundamental physical limits in resources, environmental assimilation capacity, and technological capability. It thus faces a choice between achieving a steady state in the very near future or running into a catastrophe of either man-made or natural origin that will result in a drastic reduction of present material standards. An important corollary of this view is that the world as a whole cannot possibly hope to achieve anything approaching the material living standards of the present mature industrialized countries and at the same time cannot tolerate the discrepancies in living standards that exist between the affluent "world middle class"[27] and the four-fifths of the human population that is poor. Since the pie cannot be expanded and cannot remain so unequally divided as at present, the only alternative is a drastic redistribution, which will involve radical transformation of lifestyles and aspirations in the so-called developed countries. This is usually coupled with the conviction that such a radical transformation of lifestyle would be morally superior to what we have now.[28]

The contention of the "limits to growth" school is that the people of the developed countries must curb their demands on the world's resources and their stresses on the global environment in order to leave the world's heritage for the development of the poor. Their opponents argue that the world's resource base is constantly changing and expanding due to technological progress and that, in addition, technology results in a continuous increase in the amount of consumer service per unit of resource consumption or environmental pollution. If there is any limit, it is not on the level of material consumption but on the rate at which it can be permitted to change.[29] If the rate of economic growth exceeds the rate of technological change and investment by too much the world may be in trouble. They also argue that the development of less developed countries depends in part on their ability to find markets in the developed world in order to buy the things they need for their development; consequently slowing economic growth in the developed world, far from benefiting the poor, would probably bring the whole process of development to a halt.

The practicalities of redistribution are often discussed, such as which resources would be transferred and how they would be used after transfer. Fundamentally, poverty is due not to lack of resources but to lack of productivity; individuals and groups cannot produce enough either for themselves or for exchange with others of the requirements for a decent material standard. Whether this lack of productivity is due to exploitation by the affluent or simply to the accidents of historical development and cultural evolution is also a matter of controversy. If wealth is largely due to the value added to resources by human ingenuity and skill—as in fact seems to be the case—then the problem of transfer is more difficult, because it is the skills and knowledge that are more important to transfer than the physical resources. This is why the north-south debate has recently focused on sharing of knowledge and the abolition of the concept of "intellectual property" in favor of a concept of technology and know-how as "the common heritage of mankind." In fact the same ideological dichotomy between the two concepts of technology can be found even within single countries. Some would even go as far as to suggest that technology that cannot be shared as a community heritage is not a desirable technology and should not be permitted to be deployed. The counterargument is that the ownership of technology is necessary to provide sufficient economic incentive to bring it into actual use. Here again we have an argument with both an ideological and an empirical dimension.

In the last two decades the social model for those who gave priority to equity over growth was the China of the Cultural Revolution, or rather their image of this China, which was probably quite different from the reality. In this world, prices were controlled and constant: the basic necessities of life were rationed, and economic growth was slow but steady—well below technological potential but dependable. It was a world without automobiles or personal mobility, a minimum of mechanical transport, but with basic minimum levels of food, shelter, and medical care for all. The recent political reorientation of China has cast doubt on this earlier image, both in terms of its reality and in terms of its alleged success. Clearly the Chinese people do not consider their present society as in a steady state, but instead they too, like all the other developing countries and the poor in the developed countries, are looking to something much more like the contemporary Western model as the future goal of their societies.[30] For the moment at least, they seem to be opting for growth and efficiency to some extent at the expense of equity. Even if one discounts a good deal of what has come out in the trial of the "Gang of Four," one also sees that the price of overemphasis on equity at the expense of growth and efficiency has been large excesses in the exercise of arbitrary political and bureaucratic power. The impersonal judgments of the market may not be very benign, and are often unnecessarily harsh, but societies that set egalitarian leveling as their goal also pay a large human toll. There are hints in our own experience with efforts to substitute political allocations for the market that the human costs could be high if we carried the logic of such controls to their ultimate conclusion.[31]

CENTRALIZATION VERSUS DECENTRALIZATION

The critics of modern energy technology point to the steady increase in centralization of energy production and distribution. Oil producers and refiners are among the largest corporations in the world, and their economic power is said to transcend the power of national states. Electric generating stations have steadily increased in size and become fewer in number. Distribution networks have become tied together over ever larger areas, creating greater reliability on the average but leading to the possibility of massive failures such as the two New York blackouts in 1965 and 1977. Natural gas pipelines now extend over intercontinental dimensions. Oil tankers have increased tenfold in capacity in 15 years, again with greater average reliability but with the possibility of much more massive and destructive oil spills than in the past. Natural gas is going the same way as oil, with growing numbers of liquid natural gas tankers and terminals and the large hypothetical potential for catastrophic accidents. The catastrophic potential for reactor accidents has grown more than in proportion to reactor size because in the large reactors the residual heat in case of an emergency shutdown cannot be removed by natural circulation and thus requires special emergency core cooling systems whose reliability has been called in question. The concentration of waste heat rejection from all power plants has enhanced the threat of thermal pollution from electricity generation. In short, in the view of the critics, the claimed economies of scale associated with the increasing size of energy systems have passed the point of diminishing returns because of their potential for massive environmental pollution or massive accidents and even because they are easier political targets for organized opposition to siting decisions. Furthermore, these large systems tend to separate those exposed to their risks from those who experience their benefits, thus introducing an inherent element of inequity into this type of technology.

As an alternative, say the critics, we should be looking towards the deployment of decentralized energy production systems, deliberately choosing technologies that do not have large economies of scale and that can be operated by households, small communities, or local organizations. There are both technical and political reasons for this choice. One of the most important technical reasons is that small generating plants can be configured so that both the electricity and the waste heat can be used, thus getting 80 percent of the benefit of the fuel energy rather than less than 40 percent as at present. This is called cogeneration. Among the options for cogeneration are gas turbine generators with the exhaust gases used to generate low pressure process steam, fuel cells with waste heat used for space heating, and small scale solar electric generators with waste heat captured for other uses.[32] Although cogeneration is now largely limited to oil and natural gas as fuels, it could soon be developed for use with municipal or agriculture or wood wastes as fuel, and somewhat later for coal either with fluidized bed combustion or in conjunction with a low BTU coal gasifier on-site.[33]

Other decentralized sources that have been proposed, not involving cogeneration, are wind turbines, low head "run-of-the-river" hydroelectric plants, and various forms of small scale solar heating for industrial, commercial, and residential use.

From a political point of view these sources appear attractive because they are more readily subject to local ownership and control and do not require large bureaucracies or complex government regulation. Thus, say the proponents, political and technical considerations point in the same direction, towards phasing out our centralized systems and replacing them with decentralized ones.

Does this rather attractive picture stand up to close analysis? Where are the catches? Our own belief is that the ultimate energy system is likely to be a symbiotic combination of centralized and decentralized energy sources in some kind of optimized configuration that is better than either one alone. Thus the dichotomy between centralization and decentralization may prove to be an unreal choice.

There are several reasons, both technical and political, why decentralization by itself may not be the best answer to energy production. Here are a few samples.[34]

First, in order for small decentralized systems to be affordable to consumers, they will have to be mass-produced, and will probably require a large distribution and service network. They will have to be standardized with interchangeable components. Thus the elimination of centralization could be largely illusory. The point of decentralization in the system is changed, but the fact of system centralization is not. In an electric grid the consumer has complete individual control of energy flow when he throws a switch and a flexibility that could hardly be matched by a local generator in terms of variability of load, reliability, and probably also cost. The consumer has to make a large additional investment in excess capacity to handle peak loads or else be connected to a centralized network in order to do so. If this centralized network consists of a large area system of interconnected small sources, it is difficult to see how it differs from a conventional electric grid, except that the problems of synchronization, load management, and service appear more complex and will certainly require technocratic management. The interconnected system can in fact be thought of as simply a large community generator, functionally equivalent to a large generator in a single location. It might have some advantage in degrading more gracefully in the case of large failures, but even this is not self-evident. If the energy sources are solar, wind, or hydro, the problem of matching generation to load is difficult, though not impossible, particularly if hydro is combined with either of the other two, since hydro provides storage, which can be used as a source when the other sources are not putting much out.

There is an interesting analogy between decentralized energy sources and decentralized transportation systems, which gives an insight into the problems of decentralization. By one of the criteria for decentralization the automobile appears to be an ideal device. This is particularly true of something like the old Model T Ford, which could be repaired by the individual owner and which was very simple, rugged, and cheap. The only catch was that in order to be economically viable, i.e., affordable to the average consumer, it had to be mass-produced in a highly vertically integrated factory system, and with a high degree of standardization. ("You can have any color you want as long as it's black.") Just as the automobile is the analogue of decentralized energy sources, so is public mass transportation the analogue of centralized electric grids. Mass transportation enjoys enormous economies of scale, but requires big, impersonal bureaucracies to run it, and is subject to massive breakdowns, though it is generally more reliable than the individual automobile. Unlike the central electric grid it is much less flexible in end use, since the actual service is not under the immediate control of the consumer as it is in the case of electricity or gas. It is somewhat ironic that the advocates of decentralized energy sources are often critics of the automobile and strong proponents of mass transportation.

The proponents of decentralization could, of course, argue that decentralized technologies must be pursued consistently, even if the result is higher cost. Decentralized sources do not have to be interconnected or mass-produced, and in many cases they can be manufactured locally with readily available materials, using skilled craftsmanship instead of mass-production techniques, probably at some cost in efficiency. However, it is clear that the pursuit of decentralization as a value in itself entails the sacrifice of other values, probably including that of equity because of higher costs.

Second, the automobile analogy can be used to illustrate another problem of decentralized energy sources. Because the economic viability of mass transportation depends so much on economies of scale and improves rapidly with level of patronage, it has been seriously harmed by competition from the automobile. By and large, automobile owners are more affluent than public transportation patrons, and so the proliferation of the automobile has been partially responsible for depriving the poor of an efficient and convenient form of transportation and of access to the job market. As and if decentralized energy sources become economically viable, they will probably be bought first by affluent individuals or by affluent communities. This means there will be less of a load over which to allocate the fixed costs of centralized networks, which the poor will continue to depend upon. Thus one could envision an erosion of centralized energy networks as decentralized sources are phased in, which will degrade the energy services available to the poor and to central city residents and increase costs. Whether this would actually take place is actually a very complex empirical question that has never been analyzed in sufficient detail to inspire confidence as to what would actually happen. Thus it must be stated as only a possibility, but one that clearly has serious distributional and hence ethical implications. These implications seem contrary to the values of the very people who advocate decentralized sources with the greatest enthusiasm.

Third, the economics of decentralized sources is very uncertain and will probably remain so for some time. Such sources are quite intensive in the use of materials such as steel, cement, aluminum, and glass, as well as of construction labor when this is estimated on the basis of material or labor use per unit of useful energy produced. The amount of such materials is at least several times those used in nuclear and coal plants per KW of generating capacity.[35] This seems to be more true, the more decentralized and small-scale the source being considered. Future invention may mitigate this problem, but the present situation suggests why decentralized sources are not competitive with traditional systems. In fact, the actual installed cost of solar hot water systems—the nearest to a competitive decentralized system—has escalated nearly as fast as the price of fuel oil.[36]

The reasons are not entirely clear. Some such escalation would be expected simply because the materials used require a considerable amount of oil or natural gas in their production and fabrication and because the wages of construction labor have tended to track fuel prices rather closely in the recent past. However, this cannot be the whole story. The main lesson to be drawn is that decentralized sources are not likely to become suddenly competitive with conventional sources overnight, but are more likely to be phased in gradually in a way that complements rather than replaces the existing system. This means that such sources may eventually reduce the need for new central generating capacity or for gas networks but are not likely to replace these sources. If decentralized sources are fostered through tax subsidies or other means, the additional use of resources per unit of energy produced will have adverse repercussions on the rest of the economy, since purchasing power will be diverted to energy production that could be used for other purposes. This is not necessarily bad if the social benefits of decentralization are deemed sufficiently great, but there *is* a tradeoff with competing values.

Fourth, although the economies of scale for decentralized sources, especially solar energy, are likely to be much less than for, say, nuclear power, it is not clear what the optimum size is. Moreover the geographical variability of the intensity and duration of sunlight, wind, or water flow means that the intensity of utilization of fixed capital, and hence the unit cost of energy, will have high geographical variability, certainly much more variability than in the case of nuclear power, and probably more variability even than in the case of coal, where transportation costs are more important. When seasonal effects are taken into account this variability will be even greater. For example, with solar energy the availability will be least when the requirements are highest in northern latitudes, whereas in southern latitudes where the peak load is due to air conditioning seasonal availability will tend to coincide with maximum load—a fact that will greatly amplify the unit capital cost differences among regions.

Of all energy sources nuclear power is the one that most equalizes energy costs among regions, whereas solar energy probably has the largest differentials. This is contrary to what is often implied by solar advocates, who point out that the sun is everywhere. What is neglected, of course, is that solar energy, especially decentralized solar energy, is capital–intensive, and the expensive capital has a much lower utilization factor in regions and seasons of low solar flux. This has equity implications that are not completely clear at this stage but would certainly be debated when and if solar technologies began to be economically viable on a large scale. The geographical differences also suggest that it may be more economical to generate solar electricity in regions of high insolation in centralized installations and then transmit the electricity through high voltage power lines to regions of low insolation, rather than to use solar energy in decentralized or small installations. In fact, the International Institute of Applied Systems Analysis (IIASA) energy project[37] has studied the use of solar energy for Western Europe and has concluded that the cheapest system would be to generate the energy in large central installations in the Sahara and transmit it across the Mediterranean by cable. Similar consideration may turn out to obtain for other renewable energy forms when they come to be analyzed more carefully.

The conclusion of this section is that when the question of centralization and decentralization is analyzed carefully doubt is thrown on the usual superficial comparisons. This is not to say that such comparisons are invariably wrong, but it does suggest that they may be subject to wide variations with different circumstances. Furthermore, the political advantages or attractiveness of decentralized energy generation considered by itself seems considerably less when the whole system from fuel source through generator to ultimate consumer is included in the analysis.

SIMPLE VERSUS COMPLEX TECHNOLOGIES

As we have indicated earlier one of the principal value-based criticisms of nuclear power is its esoteric character, which is contrasted with the relative simplicity and "transparency" of most renewable energy sources. In some ways this is the most persuasive argument of the nuclear critics. Although we indicated earlier that there are probably other modern technological or sociotechnical systems that are more complex than nuclear power, this would not be a telling point if in fact there were viable alternatives fulfilling the same functions that were simpler and more understandable. This may be in contrast with the international air transport system, which fulfills a social desire that cannot be met with any other technological system, irrespective of cost. Even if solar technologies or other renewable technologies prove much more costly than, say, nuclear power, they are almost certainly *technically* feasible and capable of providing a social service that is fully equivalent from the consumer's standpoint to that provided by nuclear power. If consumers dislike nuclear power enough, they can get the same benefits by paying for them, whereas in the case of air transportation they would have to compare the risks and adverse side effects such as noise with the total unavailability of high speed transportation over long distances. The same would apply to many other modern technologies with possibly adverse and differently valued disbenefits such as computers, industrial chemicals, prescription drugs, sophisticated medical technologies, or the automobile. There is thus a fundamental difference between technologies that provide an unsubstitutable societal benefit and those that merely represent alternate ways of providing equivalent societal benefits, albeit at higher cost, greater risk, or some other greater social cost.

On the other hand, there is little evidence that, outside of a small minority, the consumer really cares where his energy comes from as long as he is not directly affected by coming in contact with the production technology itself. Technical complexity or transparency are not variables that he normally considers. If he throws a switch and his appliance functions reliably he doesn't care whether the current was generated by nuclear, coal, or solar, unless the plant is close enough so that he thinks it presents a possible hazard. Even here risk rather than complexity is what concerns him. He doesn't care whether *he* can understand it provided he is confident that the people who operate the plant know what they are doing and how to deal with an emergency. The man who switches on his TV set doesn't care whether the program comes by broadcast, cable, or satellite unless this affects the quality of the picture, the variety of the programming available to him, or the cost of service. Similarly, few people who drive a car care what's under the hood unless it breaks down.

Complexity, then, is seldom an issue for the average person except via some other more directly observable characteristic such as reliability or repairability. Only if evidence of systemic breakdowns suggests to him that the technicians who operate, maintain, or repair the system are not in complete control, does he become seriously concerned. What made TMI such a traumatic event in the public mind was not the fact that anybody was hurt, even if many people did not believe the assurances of the experts that nobody was. What disturbed the public was a sense that the technology was out of control, that the people responsible didn't really understand what was going on and didn't know quite what to do. Thus, as we indicated earlier, the public's view of nuclear power is much more a question of trust in human institutions and the people responsible for them rather than any specific view of the technology itself. The major difficulty is that once doubt has been thrown on the basis for trust, however unconnected it may be to the specific characteristics of the technology itself, it is much more difficult to restore the trust than it was to create it in the first instance. It is only in this sense that complexity is an issue, at least for the overwhelming majority of the public.

CONCLUSIONS

The competition between the values of equity and efficiency constitutes an underlying theme of all the tradeoffs and policy choices discussed in the preceding sections. In response to rapidly changing price structures, price controls, in combination with various forms of rationing or political allocation of scarce supplies, are proposed to preserve social equity and to soften the hardships of economic adjustments to high energy prices. Yet we have seen that such a policy requires increasingly detailed government intervention in the decisions of individuals and businesses, which are difficult to implement in a fair way and create a bureaucracy that is both inefficient and unresponsive. Furthermore, the choice of political allocation as the mode of allocating scarcity opens the whole political process to an intense competition for special advantage among competing interests, with often quite unforeseen effects on equity. In addition, regulated prices have resulted in overconsumption and generated a demand for increased oil imports contrary to the original objectives of U.S. energy policy, and contrary to our foreign policy objectives. Moreover, American demand has driven up world oil prices, increased the vulnerability of the world oil system to further political disruption and cartelization, and transferred American adjustment costs to other countries, particularly to the nations of the Third World that could least afford it. Thus striving for domestic equity has exacerbated some international inequities and made the American economy itself more susceptible to subsequent price jumps, so that it is doubtful whether even equity has been served in the long run.

Similarly, in the debate between centralization and decentralization in the production of energy the effects on equity in practice are ambiguous. The advantage of decentralization is that the risks and costs of energy production are borne more fully by those who benefit from energy, which is seen as more equitable. At the same time, we have also seen that decentralization has the potential for favoring more affluent groups and regions by increasing the geographical disparities in energy costs and eroding the economies of scale in existing energy grids. Both equity and efficiency seem to call for increasing interconnections in the decentralized system, thus throwing into question the original alleged advantages of decentralization. Proponents of decentralization sometimes argue for them in the name of efficiency also, suggesting that higher capital costs may be offset by savings in overhead for managing and controlling complex systems; this, however, is an empirical question rather than a value question, which depends on the details of design and implementation.

Similarly, in the example of simple vs. complex technologies there appears to be a tradeoff between transparency at the point of consumption and complexity at the point of production. Focusing only on simplicity in production technology and conversion technology without examining details of end use may be completely misleading with respect to the underlying social value issues involved. Again the final answer will depend on details of design and implementation, which are difficult to sort out by looking only at the broad characteristics of particular production technologies.

The lesson is that the choice of energy technologies should not be made solely on the basis of arguments over abstract sociopolitical values—arguments that take no account of what is likely to happen when such choices are actually implemented. As we have seen, realistic assessments of the practical implications often disclose unanticipated complexities and unintended consequences. The same empirical test must also be used to identify the risks posed by various technological options, as we shall try to show in the next chapter.

7 Values and Policies: Dilemmas in Risk Assessment

INTRODUCTION

In the last chapter we considered energy policy choices largely in terms of social and political considerations. In this chapter we look at the problem of risk assessment for various energy technologies and try to show that the value choices implicit in assigning various weights to these risks once again depend to a greater degree on the details of design and implementation than is apparent in generalized discussions of the issue.

ECOLOGY VERSUS HEALTH/SAFETY

There is probably no subject more confused in the public mind than the difference between ecological effects and health/safety effects. From a strictly ecological viewpoint nuclear power is almost certainly the most benign of all the main available or prospective energy supply technologies. The total amount of land that has to be disturbed in connection with the nuclear fuel cycle, for example, is generally less than for other systems, even after taking into account the land required for uranium mining and waste disposal.[1] If breeders become widely deployed, the uranium mining component of land use would practically disappear, and ultimately the land used for waste disposal may be reduced because of the possibility of recycling the long-lived isotopes through the fuel cycle (although this is problematical at present, and a long time in the future in any case). Thermal pollution is an ecological problem and is about 50 percent greater per unit of nuclear electricity produced than for other electricity generating sources. However, this is true only with the present commercial type of light-water reactor. With more advanced reactor types of the future, including breeders, the thermal pollution would be no worse than for other means of electricity generation.

Only with the widespread use of cogeneration or district heating would the thermal pollution per unit of useful energy be significantly reduced relative to current technologies. Indeed, the possibility of making direct use of waste heat locally represents one of the most significant potential technological advantages of decentralized forms of power production. Protection of human health is so much more stringent a criterion for reactor operation, accident avoidance, and waste disposal than is protection of the ecosystem that the latter never has to be considered if the former is satisfied. This is because in ecosystems we are concerned only about survival of populations, whereas with health/safety we are concerned about individuals.

By contrast, hydroelectric dams probably represent the energy source most threatening to natural ecosystems per unit of energy produced. This is because of the large amount of natural habitat that is affected by flooding and the effects of siltation above the dam and absence of siltation below the dam.[2] Of course one cannot completely generalize this statement because the ecological effects of dams are so location-specific. On the other hand, in North America the dam sites that are least ecologically damaging have mostly been used, so that if we tried to expand hydro significantly we would rapidly run into more and more ecologically damaging sites.

Coal presents both ecological and health effects because of the ecological effects of mining and of acid rain and the potential health effects of air pollution and the high accident rate in underground coal mining and transportation by rail.

One reason these comparisons present a dilemma of values is that, while people tend to be in general agreement about the degree of avoidance of health damage that is desirable, they have widely differing assessments of the undesirability of ecosystem damage. Some of this may be due simply to technical ignorance. For example, the average person, even if he is fairly well informed technically, may not be aware of the long-term adverse human effects of the loss of genetic diversity, a prime effect of loss of habitat. On the other hand, the destruction of wild flowing rivers or backpacking areas, or pristine wilderness areas is much more a matter of aesthetics and personal values and differs widely among different evaluators. The variability of understanding and values poses a genuine dilemma for the choice between hydro and nuclear particularly, as has been well illustrated by the energy debate in Sweden, which is well endowed with potential for both hydro and nuclear.

Solar technologies, if deployed on a very large scale in centralized installations, e.g., in the southwestern desert areas, could present serious ecological problems; the same would be true of the massive deployment of wind generators. For example, wind generators would have to be interconnected, and they would require a dense network of access roads for maintenance. Like other solar technologies they use large amounts of energy intensive materials whose extraction and processing would be a secondary source of ecosystem damage. It is very difficult to quantify such effects, since nobody can agree as to just what should really be counted against the energy system.[3]

The value dilemma in all this arises from the relative weight that people are prepared to put on remote but real health risks in comparison with risks to more intangible values such as wild habitat, genetic diversity, atmospheric visibility in pristine areas, endangered species, or disturbance of hydrological systems. The discussion is made more difficult by the fact that there are large uncertainties associated with new and untried technologies with which we do not have practical experience on a large scale. Often the risks of a technology depend less on its general characteristics than on small practical details of its implementation. Thus, technologies that appear benign in principle may turn out to be quite damaging in practice, unless they are carefully designed with their secondary impacts in mind. The question can be asked as to whether the problematic aspects can be recognized sufficiently early in the development of an industry so that technical or managerial fixes are not too expensive or difficult. The answer depends on whether technological assessments yield valid and practical results and whether they are kept current with research carried on in parallel with the development and deployment of the technology itself.

"RICHER IS SAFER"

Historically, there has been an empirical correlation between the general level of health and safety and the material standard of living. This has been true both over time in a given country and between different income classes and regions. The correlation is far from being 100 percent, but the trend is unmistakable. In about the last decade some theoretical doubts have been cast on the implications of this correlation because of the rise of concern with the hazards of low-level toxic materials, which exhibit long time delays beween exposure, and the appearance of detectable health effects. In theory one could accumulate large dose commitments with no detectable adverse effects, only to find 30 or 40 years later an epidemic of cancers or other health damage sufficient to reverse the improving trend in overall health statistics. So far such delayed effects have only appeared in very limited or localized populations, such as residents near hazardous chemical waste dumps, asbestos workers, chemical workers exposed to vinyl chloride monomer, textile workers exposed to cotton dust ("brown lung"), or coal miners exposed to coal dust ("black lung"). Occasional alarms have been sounded with respect to radiation workers (Hanford plant workers, Portsmouth Navy Yard workers, residents near the Rocky Flats plutonium plant), but in almost every case reanalysis of the data by other scientists has discredited the original work.[4] In none of these examples has the exposed population been large enough to affect the gross health statistics of a whole region. To date average life expectation in all the industrialized countries except the Soviet Union has continued to increase with increasing material standards and energy consumption, albeit at a declining rate. Nevertheless, because of the existence of health effects with long delay times one must say that historical experience is not an infallible guide to the future. The large geographical differences in the incidence of different types of cancer even among regions with about the same living standard suggests that environmental (including dietary) factors are important, although we have no reason to say that most of these environmental factors are man-made in origin, with the single exception of smoking.[5]

Thus, subject to the caveat outlined in the preceding paragraph, it does seem to be broadly true that "richer is safer,"[6] and this suggests the theoretical possibility that there is a point of diminishing returns in reducing the risks associated with a host of individual human activities or technologies. If the cumulative impact of these risk reduction efforts is a significant reduction in material living standards at some future time, it becomes uncertain whether the risk reduction efforts were worthwhile. Today we simply do not have the knowledge to say where this balance point lies. There is also another consideration. To the extent that risk reduction requires the addition of safety devices and equipment to existing equipment, the risks associated with the production and installation of such equipment could eventually outweigh the risks eliminated by it.[7] Again this is a point we do not know enough to evaluate practically. We should note, however, a belief among many nuclear engineers that we may have already passed this balance point in relation to nuclear reactors where additional back-up devices may make them less safe. The typical problem is that devices to reduce the consequences of very improbable events may increase the probability or consequences of more probable though less consequential events. Another example is that risk reduction efforts that may have a positive benefit relative to cost in an industrial country may have negative benefit relative to cost in a poor country. An example may be controls on the distribution of certain kinds of contraceptive devices or drugs.

The value issue that is raised by these considerations relates to the effort and resources that society ought to devote to risk reduction. The cumulative effect of large numbers of risk reduction efforts, each one apparently beneficial when considered in isolation, may not reduce the risks across society as a whole, but could actually increase them.

FUTURE RISK VERSUS TECHNOLOGICAL CAPACITY

There is another issue that is closely related to the "richer is safer" question described in the preceding section. If economic growth and technical progress continue at a faster rate our descendants will have greater capacity to deal with the environmental and resource depletion problems that we bequeath to them as a result of current activities. One cannot make a prima facie case either way as to whether the gain in future capacity to solve problems will offset the increased severity of the problems our descendants may inherit. Clearly, to the extent that we can reduce future problems for our descendants without compromising their problem solving capacity we are ahead of the game, and we owe at least this much to future generations. But it is not obvious how much more than this we owe. If safety and health were costless, or if all the effects of our current actions and future remedies for them were foreseeable with certainty, our ethical obligations would be clear. But Albert Hirschmann's "Principle of the Hiding Hand"[8] puts a very different light on this question and changes an ethical imperative into a policy puzzle. Hirschmann's point is that most social or entrepreneurial initiatives have to be taken without full knowledge of their consequences. Bad consequences and difficulties often appear that were not anticipated, but these are frequently compensated by unexpected discoveries and technical opportunities that could not be foreseen until one was well embarked on the activity. Thus, there is a serious danger in trying to exercise too much foresight, going too far beyond what can be fairly rigorously inferred from what we now know; there is a case for relying on a little bit of luck, which is what Hirschmann really means by his "hiding hand."

Even if current risk reduction efforts on behalf of our descendants were costless, it is important to take into account future technical capabilities that will result from scientific progress. Two hypothetical examples will serve to illustrate our meaning. The main health effect of ionizing radiation is, of course, the increased risk of cancer, with genetic damage to future generations being a secondary and less certain effect. Yet, it is quite conceivable, perhaps even likely in the minds of some, that future scientific advances will lead to means for curing such cancers, or for neutralizing induced cancer risk from radiation through simple preventive or immediate postexposure therapeutic measures. If we were certain that this would happen we might justifiably be much less worried about the safe long-term storage of radioactive wastes. The practical question, which is really an ethical question, is: Just how much weight in present decisions should be given the likelihood of our being able to neutralize the deleterious effects of radiation at some time in the future? Probably most people today would say we should ignore the possibility, but this by itself represents an implicit value judgment that might not have been shared by our ancestors or agreed to by our descendants. The only point that must be made here is that it is a technical consideration, which introduces an additional element of ethical ambiguity into the discussion of intergenerational obligations.

One could also analyze the same issue in terms of discounting. When we are talking about economic benefits or costs it is general practice to use some rate of discount to compare future benefit or cost streams with present costs. Future benefits are less valuable than present benefits, the more so the further they are in the future. Yet, when we talk about long-term health risks, such as those from exposure to radiation or carcinogenic chemicals, the tendency is to assign the same weight to a future fatality on grounds of the psychological trauma of knowing one is at increased risk of contracting a dread disease as a result of a discrete current event. We could also argue, however, that delayed fatalities ought to be weighted with some sort of discount rate tied to the cumulative probability of the discovery of mitigating or preventive measures over the period of time delay. The Risk/Impact Panel of CONAES has pointed out that on the average the survival rate for diagnosed cancers has been improving at the rate of about 0.5 percent per year over the last 40 years.[9] Should this information be incorporated into a discount rate to be applied to future fatalities arising from radiation exposures that are the result of current activities? Of course, the argument in its present form is suspect, for the survival figure is a composite among many different types of cancer and not necessarily those most frequently connected with exposure to radiation or industrial chemicals. Yet, even though this argument may be of dubious validity in its current form, it embodies a principle that may be valid, or is at least arguable.

The second example is that of the CO_2 effect on climate associated with fossil fuel combustion. Even if the consumption of fossil fuels should continue to increase worldwide at pre-1973 rates, climatic effects would not be major in all probability before 2020 at the earliest. More likely, the problems will not be acute until closer to the middle of the twenty-first century. By that time agricultural technology may have advanced to the point where adaptation to extensive world climatic changes would be quite feasible. Indeed, it is conceivable that the food production system will have become completely industrialized with the aid of recombinant DNA technology and closed system recycling of essential nutrients in climatically controlled spaces.[10] In all probability this happy situation might come about sooner if economic growth is faster, which is likely to be the case in the absence of constraints on the use of fossil fuels. These future technical possibilities generate the same sort of ethical ambiguity that we cited in the case of long-term radiation effects. The implication is that we may be about to take the CO_2 problem too seriously because of our failure to take into account future technical progress. On the other hand, such technical progress is by no means assured, and the question is how much we can or should count on it in our current decisions. The same type of dilemma arises in reverse when it is argued that we should forego coal or nuclear generated electricity because solar technologies can come along to rescue us in time from the depletion of oil and gas. These parallel examples are interesting because different sides of the argument rest a part of their case on technological optimism. In one case the optimism is with respect to our future enhanced capacity to fix the problems we generate today. In the other it pertains to our future ability to deploy more benign technologies than those available to us today, and thus bypass currently available options to which we would prefer not to commit ourselves.

A third example will serve to illustrate the same type of dilemma in a different way. Future research is very likely to lead to safer and more certain ways of permanently disposing of radioactive wastes, especially high level wastes in the form of spent fuel or reprocessing wastes from nuclear fuel cycles. On the other hand, once waste is disposed of irretrievably it may pose a risk to future generations that could have been greatly reduced had we waited longer until we found a still better method. The risks of temporary storage in the meantime might have been large, but incurred over a time period small compared with the eons over which the risk is incurred once an irreversible commitment to a particular method of storage and a particular site has been made. There is a two-way tradeoff between temporary storage and early permanent storage, but there is a third element determined by how much waste we permit to accumulate before we should insist on a permanent rather than a temporary (retrievable) solution. Producing less waste may in practice just remove pressures for finding a permanent solution, and because we already have a large accumulation from both civilian and military nuclear programs worldwide, we cannot afford to ignore the problem even if we were to abandon nuclear power altogether. Taking into account political realities, are we more likely to find the safest possible solution for existing wastes if we continue with nuclear power growth than if we do not? Given the complex technical tradeoffs that actually exist, ethical principles related to our obligation to future generations give us very little practical guidance as to how we should decide.

COMPARISON OF LONG-TERM THREATS

There are many long-term health and safety threats associated with both the production and consumption of energy. In a certain sense even the annual automobile fatality toll can be viewed as a consequence of the *consumption* of petroleum products. Since annual energy use for automobiles is about 20 quads and there are about 50,000 annual fatalities, this amounts to about 2,500 fatalities per quad. (A quad = 1 quadrillion [10^{15}] British Thermal Units [BTUs] of energy.) For a coal burning electric power plant the corresponding figure comes to about 100 fatalities per quad, taking into account accidents only. For nuclear power the corresponding estimate for accidents only is roughly 2 per quad. For the population served by nuclear reactors generating a quad per year the cancer death rate from all other causes is about seven million. The questions suggested by these numerical examples are: What fatality rate should society be willing to accept for these various activities? What basis should be used to arrive at a *relative* standard of societally acceptable fatality rates for different technologies? Or, how much should society spend to avoid one additional fatality associated with a given technology?

In this last question there is a tacit implication, often asserted explicitly by economists, that society should set standards so as to equalize the cost per fatality avoided. This would be especially true in comparing alternate technologies that produce approximately the same societal benefit, e.g., alternate means for electric power generation. Yet, there are deep ethical questions buried in this judgment. One can cite extreme hypothetical examples in which such a principle of equal cost per death avoided would be seen to be highly inequitable. Suppose that in a nuclear plant one had to send in one worker a year who would receive a fatal radiation exposure, and this could be avoided by the expenditure of an extra $5 million in plant construction. Suppose, on the other hand, that routine emissions from a nuclear plant exposed a population of a million people to very low levels of radiation and that this could be reduced enough to save one cancer death per year at a cost of $5 million in reactor construction (this is the actual standard set by EPA for routine emissions). Would these be equitable and comparable criteria? Clearly not. There is a difference between producing one extra fatality among a million people, 350,000 of whom will die of cancer anyway, and one identified person who is virtually certain to die of cancer. We would probably spend a lot more, or forego more valuable benefits, to avoid a certain death than we would to avoid one statistical death in a large population subject to many other risks.

The question concealed beneath this extreme example is whether our criterion for setting health standards should be based on equal risk to individuals or on equal population risks. In radiation standards, the criterion is usually population exposure, but we also set standards for the maximum exposure of any single member of the public or any single nuclear worker. Yet, if the standards applied to a single worker were permitted to a large population, the number of fatalities would be unacceptably large. The point is here that in setting radiation exposure standards one implicitly considers both risks to individuals and population risks and the relative weight accorded to these criteria is a value judgment that cannot be given an objective basis. Even if one used as a sole criterion maximum risk to an individual, there is a factor of 1,000 difference between the risk permitted to a member of the public at the site boundary of a reactor installation and the maximum risk permitted to an individual worker in the plant.

The difference is fundamentally arbitrary, although one might argue that the plant worker's risk is voluntary whereas that of a member of the public on the site boundary is involuntary, and there is some evidence on actual risks accepted by society that indicates people will tolerate about 1,000 times greater risk in activities that are clearly undertaken voluntarily.[11]

When it comes to comparing widely different technologies the differences in standards become even more spectacular. For example, the EPA requirement that routine emissions from reactors or fuel cycles should be reduced to the point where the marginal cost of avoiding a person-rem of radiation exposure is $1,000. Using the linear dose-response assumption and the value of 2×10^{-4} cancers per person-rem, this corresponds to an investment of $5 million per additional life saved. By contrast, actual practice for highway safety improvements seems to correspond to about $50,000 per fatality avoided, while internal safety features in the car correspond to about $250,000 per fatality avoided.[12] Various rationales have been given for these differences, including the difference between voluntary and involuntary risks and the fact that higher benefits justify higher risks. But these matters are highly judgmental and subjective.

In addition to the question of comparing cost per unit of risk avoidance among different technologies, there is also the issue of uncertainty. In the literature risks are usually quoted in terms of expectation values of harm or of probabilities for a given level of harm. There are usually large uncertainties or error bands both in the estimation of harm from a given event and in the probability of particular events. In regard to the famous reactor safety study (Rasmussen Report)[13] there is now a general consensus that the estimates of probability of the most severe accidents could be too high or too low by a factor of the order of 100, a total range of about 10,000 factor.[14] What this means is that it is extremely unlikely but not impossible that the estimates are off in either direction by this large a factor.

How should these facts be used in judging whether a reactor is safe enough, or—even more difficult—whether a future nuclear industry will be safe enough, taking into account *both* the future growth in total nuclear capacity and the likely improvements in both safety itself and in the range of uncertainty of the probability of risk? These are ethical judgments that boil down to a concern with where the burden of proof should lie in relation to the introduction of a new technology for which estimates of the probability of harm are necessarily hypothetical and based on theoretical models that, by their inherent nature, cannot be empirically validated in relation to every element and assumption. These considerations are further complicated by the observation that an event with an estimated frequency of once in 20,000 years (the estimate for a reactor blowdown with failure of emergency core cooling) could just as likely occur next month as 20,000 years from now. The estimate does *not* mean, as often intuitively thought by a layman, that we can relax and wait for 20,000 years before we have to worry about a meltdown.

Another example of the dilemma in setting standards or evaluating potential events is provided by a detailed comparison between the danger of reactor meltdown and the radioactivity hazards arising from conservation measures. One widely recommended conservation measure is to reduce the leakage of air from residential structures by a factor of two. Reducing leakage results in the buildup of the radioactive gas radon, which is continuously emitted from the ground or from cellar walls in the house and builds up a higher concentration if the air is changed less frequently.[15] If these particular conservation measures were to be applied almost universally with no provision for a special air exchanger, which also exchanged heat, the resulting population exposure to radioactivity would be considerable and, indeed, would be equivalent to the population exposure that would result from a major reactor meltdown once a year, excluding the few people very near the reactor site. In fact, it might be asked whether there is not a fundamental inconsistency between the enormous concern with regulation of the small amounts of radioactivity from the nuclear industry and the exposures of the whole population just resulting from the fact that people spend most of their lives in buildings. This question is especially relevant in view of the fact that there are probably several rather simple technical measures that could be used to reduce the radon exposure in buildings and that would cost much less per person-rem of exposure avoided than the $1,000 per person-rem specified by EPA for the nuclear industry (but not for medical xrays).

NUCLEAR POWER VERSUS NUCLEAR WAR

There are widely differing opinions as to the degree to which expansion of nuclear power could reduce our dependence on foreign oil. There is little question, however, that if the plans for nuclear expansion that existed in the early 1970s would be realized in the rest of the twentieth century, and if we had followed a policy of maximum electrification of the economy on the consumption side, substituting electricity for fluid fuels where it appeared close to cost-effective, the United States could have become independent of oil imports well before the year 2000. Since the United States is already the largest oil importer in the world, and hence has a very large impact on the world oil market and on international competition for oil, our achievement of petroleum independence could have greatly moderated world competition for oil. The question that this raises, then, is whether the moderation of this competition would have sufficiently reduced the chance of conflict, which might ultimately escalate to nuclear war, to offset the increased chance of accidental nuclear war arising from the proliferation of nuclear know-how, and hence possibly nuclear weapons, to more and more nations. This is a very speculative question, which cannot be answered with any confidence, but it is certainly relevant to consideration of the dangers of proliferation arising from the diffusion of nuclear technology and nuclear materials.[16]

The damage that could result from a nuclear exchange so far exceeds the damage that could result from a nuclear reactor accident, or even from sabotage of a nuclear reactor, that even a minuscule increase in the possibility of nuclear conflict would be enough to offset any dangers of nuclear power. To put the matter in converse form, if the more rapid deployment of nuclear power would have even a tiny impact on the possibility of nuclear war, either positive or negative, this would be much more important in human terms than any direct health or safety threats connected with civilian technology. Yet, this is an issue that is seldom addressed, perhaps in part because it is almost impossible to come to grips with scientifically, whereas reactor safety and waste disposal and even the early stages of weapons proliferation are much easier to discuss technically. Indeed, in the whole nuclear power debate we may be falling into the trap of the "technical fix," arguing over technical issues of minor importance because they are easy to describe fairly precisely while we avoid highly uncertain speculations about future human behavior.

There is another more subtle issue involved here, which relates more to internal than to external politics. This has to do with the effect of the deployment of nuclear power on the democratic process. Here the concern of the critics is not directly with the risks of nuclear power as such but with the political dynamics of preventing sabotage, theft of fissionable material, or other malevolent abuse of peaceful nuclear technology. The contention is that protection of society against the abuse of nuclear technology will by itself inevitably lead to subversion of the democratic process, and that, no matter how low the actual risks against which protection is sought, the threat to democratic institutions makes the game not worth the candle. A single threatening incident, or even an alleged or hypothesized incident, could lead to a panicky political overreaction, which could threaten civil rights and due process to an unacceptable degree. The argument would be that such threats would constitute a much greater and more imminent danger to society than any physical dangers and should not be risked. This kind of argument is as difficult to evaluate as those involving the risk of nuclear war because it rests on the same sort of speculations about political behavior. In fact, such arguments almost can be considered to rest on assumptions about the nature of society, which are impossible to subject to empirical examination, and so are not likely to be resolved by rational argument alone.

RENEWABLE VERSUS NONRENEWABLE RESOURCES

The term renewable resources has come to acquire benign moral overtones that, in our opinion, have tended to confuse the energy debate. Part of the problem is that the term "renewable" implies an absolute property that can only be approximated in practice. Renewable energy forms, in order to be made usefully available, require an investment of nonrenewable resources, including the use of nonrenewable fossil energy to extract, fabricate, or recycle the materials used in devices for making renewable energy available. Even biomass, if harvested on a self-sustaining basis, requires the use of nonrenewable nutrients, such as phosphorus, which probably cannot be completely recycled through the system. On the other side of the coin, there are many resources that are nominally nonrenewable that are so abundant in the earth's crust that to classify them as nonrenewable is essentially academic. This is true, for example, of natural uranium when used as a feed for breeder reactors. It may be partially true of some heavy oils and oil shales, which are much more abundant than conventional oil and natural gas. It could be true of coal on a medium time scale, especially if it proves possible to gasify otherwise inaccessible coal beds *in situ* underground. The point is that renewability and nonrenewability are really quantitative concepts. The question is what quantity of exhaustible resources is used to produce a given quantity of useful final energy, taking into account the fact that no capital installation lasts forever or can be recycled 100 percent. There is no special virtue to renewable resources except in this limited quantitative sense.

Another concept often used in discussions of renewable energy sources is that of stocks and flows. Solar electricity and natural gas represent two extremes in a spectrum of energy sources in which there is wide variation in the ratio of nonrenewable resources that go into fixed capital as compared with expendable fuel. The "fuel" in a solar installation is a free good in infinite supply; the entire consumption of nonrenewable resources is in investment. If the energy source is capital-intensive, as solar electricity is, and if the particular energy source is to fill a growing market, as it is bound to do if it is a source being newly introduced, then the consumption of nonrenewable resources will depend sensitively on the rate of buildup of the new source. If the nonrenewable resources used in the buildup require the expenditure of nonrenewable energy forms in their manufacture, then the question of whether there is any net energy output from the system becomes relevant. If the source, which is a "stock" source, is economically competitive with other "flow" energy sources such as natural gas, then net energy is of little concern because favorable economics guarantees favorable net energy. But if it is decided to subsidize the large scale deployment of a particular energy source despite its higher economic cost than other sources, then there is a possibility that the nonrenewable energy consumed in the buildup of the new source could exceed the new energy it would produce for as long as the buildup period continues.

This is the reason why the rapid buildup of a subsidized solar industry could reduce the net availability of fossil fuels during the buildup period, which might be 20 or 30 years if the source were designed to replace existing "flow" sources. There is a case, of course, for investing in fossil energy while it is cheap in order to produce solar electricity to be available at a time when the fossil energy is much more expensive. In this sense solar installations, just like enriched uranium for nuclear fuel, could be looked upon as a way of stockpiling fossil fuel as insurance against future shortages. It then becomes an economic question how this form of stockpiling compares with more conventional stockpiling such as a strategic petroleum reserve. One could think of using imported oil while it is available in order to create enriched uranium or solar installations to be used in case of an oil cutoff. We have not evaluated this option compared with filling the reserve. It would probably be worthwhile only in the face of very long supply interruptions, if ever.

One could also make a moral argument that the technologically capable nations ought to use their ingenuity and skills to build capital-intensive energy production installations in order to free cheaper nonrenewable energy resources for use by developing nations while they are still relatively abundant, but of course such stockpiling of technology could place a temporary surge of demand on world oil and gas supplies, which may make them less available to developing nations in the near term. Thus there is a fine balance here. While the bottom line is ultimately an ethical judgment, the final outcome still depends on the result of a complex and subtle technical-economic analysis. It is impossible to arrive at a correct ethical answer without great technical sophistication. Thus, while the ethical principle at stake is quite a simple and understandable one, the practical policy consequences of its correct implementation cannot be inferred from ethical reasoning by itself. We believe this is a typical rather than an unusual example of the problems of applying ethical principles to the choice of energy policy.

CONCLUSIONS

Risk assessment has always been central to the public discussion over choice of energy production technologies, yet its application in practice leads to many ambiguities in the applicability of ethical principles. We must recognize at the outset that all methods of energy production pose some risks to health, safety, or the environment, but that more severe risks of this type are often associated with technologies for the end use of energy than with technologies for its production, conversion, or distribution. This is complicated, however, by the fact that the risks of consumption technologies are frequently at least quasi-voluntary, whereas the risks of production, conversion, or distribution are frequently imposed by society involuntarily on those affected. From the standpoint of overall societal benefit one could make a strong case for devoting more social resources to the assessment and reduction of risks associated with end use, but this may run counter to other values of society relative to the imposition of involuntary risks; once again the equity versus efficiency issue raises its head in a new guise.

Technology that may pose long-term threats to health and safety, such as nuclear power, may be especially benign from a strictly ecological standpoint, and the reverse may also be true as in the case of hydroelectric power and certain types of renewable technologies. This poses difficult choices because ecological risks are more subtle and less immediate or visible in terms of their ultimate human consequences, making value tradeoffs somewhat incommensurable with each other. Individuals and groups, moreover, differ much more widely in their sensitivity to and valuation of ecological effects than health and safety risks.

There is also a closely related question of how much risk avoidance for the immediate future may affect our capacity to deal with risks in the longer term, and how this should be used to discount future hazards. In particular this involves weighing the general improvement in health and safety, which has historically accompanied the improvement in material welfare and economic growth, against the desire to eliminate or avoid presently perceived risks. The problem of value tradeoff is made more difficult by the inevitably speculative nature of future improvements.

There is also a tendency to focus on risks that can be more or less quantified or demonstrated in the here and now, even though they may be small, while losing sight of risks whose estimation is much more speculative, often because it depends upon judgments about human behavior and the future evolution of human institutions, value preferences, and managerial capabilities. Here the effect of the deployment of nuclear power on the future probability of nuclear war or the ultimate elimination of nuclear weapons from the world's arsenals poses a specially difficult problem. This is because only a miniscule change, either positive or negative, in the probability of nuclear war would completely swamp in its ultimate human implications any other technology-based hazard about which we debate intensely.

Similarly, the debate between the advocates of renewable and nonrenewable energy resources may appear very different when posed in general terms than when posed in relation to the practicality of implementation of specific technologies. In part this is true because no renewable resource can be delivered to end use without drawing on some nonrenewable resources, while certain nonrenewable resources are so abundant when used with certain technologies that some of their social characteristics are indistinguishable from those of renewable technologies.

8 Energy and Democracy: Values and Process

INTRODUCTION: THE POLITICAL STAKES

In one respect, Americans are less seriously affected by the energy problem than people in other countries: our physical habitat consists of a large part of a continent comparatively well endowed with fossil fuels, and we have the capital and technological capacities necessary to explore for these resources and to develop alternate sources. Yet, despite our abundant resources, we are seriously affected by the energy problem in other respects. Because energy supplies have been especially plentiful (even in earlier times when Americans could afford to be profligate in their use of wood), our economy and our lifestyle have become increasingly energy-intensive. We consume far more energy per capita than people in almost any other country. As has been noted earlier, Americans comprise only about 5 percent of the world's population but account for about a third of all the energy consumed by the four billion people who currently live on our planet. In our homes, industries, and vehicles, we use about 50 percent more energy than Europeans use, relative to the level of economic production. Only Canadians—with a generally less temperate climate, a lower ratio of people to territorial space, and an even more energy-intensive industrial mix—consume more on the average than we do.[1] As a result of this high and—until recently—growing dependency, our economy and our way of life have become as vulnerable as any to disruptions of supply and to the problems entailed by efforts to assure and expand sources of supply.

Our energy problem is acute not only because we have come to be so dependent upon supplies of energy, but also because most of the primary energy we now use is in the form either of oil or of natural gas. These two sources now serve as fuel for 75 percent of all the energy services we use. Domestic supplies of both these "premium" fuels (so called because of their versatility and relative "cleanliness") have declined in recent years because of the exhaustion of the most readily accessible sources and because government policies encouraged reliance on imports and discouraged domestic exploration, especially of natural gas. As a result, we have become dependent on imported oil. Because our appetite for oil has been matched by even more steeply rising world demand, the oil exporting countries have been able to take advantage of their control over this scarce resource by behaving as a quasi-cartel, by attempting to fix prices and control supply in order to maximize returns. All countries dependent upon oil imports, not only for fuel but as feedstock for use in fertilizer, plastics, and other petrochemical products, have been jolted by steeply rising prices.

The oil-consuming nations have also become vulnerable to major economic and social dislocations due to unpredictable shortages and interruptions of supply. They are all the more vulnerable because, geographically, the largest supply of proven reserves of oil is located in a remote and politically volatile region, linked to the consuming countries by long oceanic supply lines subject to interdiction at various points. Prolonged embargoes or interruptions of supplies due to the use of the "oil weapon," or to local conflicts of various kinds, could cripple the economies of the great industrial states upon which the world depends both for food and manufactured goods and as markets. Major disruptions could trigger massive unemployment and deep depressions, which, in turn, could generate demands for the use of military force to secure oil supplies. If the two rival world power blocs should become competitors for resources vital to both of them, one of the most plausible scenarios for nuclear confrontation would present itself.

In the long run—variously estimated at between 20 and a hundred years—the more affluent industrial countries should be able to diminish their reliance on fossil fuels, especially oil and natural gas, and move toward a much greater usage of alternative sources, which for practical purposes could be considered inexhaustible—though other considerations (such as the effect on the environment) might well impose limitations on the use of such energy. The alternatives currently envisioned include the various forms of solar energy, nuclear fusion, and the breeder reactor. In the near term, the exploitation of nuclear power, in facilities in being and under construction, together with the fuller exploitation of the world's prodigious resources of coal and, in certain regions, of geothermal energy, should make possible a successful transition to a pattern of consumption considerably less dependent upon oil and natural gas. It will take time, however, to develop these alternate technologies and to build the facilities and equipment necessary to exploit them economically.

In the meantime, the energy problem will remain a leading issue on the agendas of domestic and foreign policy. Unfortunately, the very persistence of the problem in a form that is only intermittently acute makes it unlike other political crises, which command attention and resources until solutions are found. Wars and depressions, for example, tend to have more concentrated and visible impact. It is harder to focus national will and to formulate a single national policy to cope with a problem that has so many different facets, whose salience is for the most part indirectly felt, and whose impact varies over time and may be felt more acutely in the future. The energy problem is a critical component of more visible phenomena such as inflation, unemployment, and the political assertiveness and instability of oil-rich countries, but it will not be resolved once and for all by a single set of decisions. As Brooks and Hollander have explained:

> The energy problem is not a crisis in the traditional sense, analogous to a military national emergency. It is, rather, a gradually deteriorating situation that could erupt into a crisis at any time, nonetheless a situation on which it is difficult to focus sustained public attention because the symptoms of the crisis appear only sporadically and unpredictably, as in the 1973 oil embargo, the 1977 US natural gas shortage, and the 1979 political upheaval in Iran.[2]

To this list can already be added another episode: the 1980 cutoff of oil supplies from both Iran and Iraq as a result of the war between these two countries and the subsequent steep rise in OPEC prices. These episodes have made it clear, however, that the energy problem cannot be dismissed as a temporary aberration, due entirely to the use of the oil weapon by the Arab producing states during the Yom Kippur War, or as a result of the formation of an economic cartel that is bound to become unravelled as demand for oil drops due to increased prices. Even if world consumption should decline from the present 60 million barrels a day to as low as 40 million barrels a day by the 1990s, the general cost of oil and other forms of energy will remain high and the interim period will remain one of great vulnerability.

After an initial period of skepticism, the increased prices have aroused a general awareness of the seriousness and long-term nature of the energy problem. American consumers have curbed their consumption significantly and have become acutely aware of operating costs in their purchases of such energy-intensive items as appliances, automobiles, and housing. Just as significantly, rising costs have already led to dramatic improvements in energy efficiency by industrial producers, and more are in prospect as the necessary capital expenditures and technological improvements continue to be made. Industrial conservation is reported to have already achieved a reduction of 14 percent per unit of output since 1973.[3] Shifts by large users, particularly electricity-generating utilities, from oil and gas to coal are also in prospect.

The pace of change has so far been slow both because natural gas supplies have been more readily available than anticipated and because of difficulties in getting more coal out of the mines and into assured systems of transportation after decades of decline in that sector. Major initiatives have been taken, however, aimed at reducing dependency on imported oil, stimulating the development of alternate sources, providing incentives for conservation and improved energy end-use efficiency, and assisting utilities in obtaining access to investment capital. Underlying these efforts is the belief that the best strategy for dealing with the energy problem in the near term is to maintain accustomed levels of energy services by making capital investments that will permit reductions in the use of fuel. If this strategy proves feasible, continually rising oil prices will not cause significant inflation and thereby prevent the U.S. economy from attaining a reasonable rate of growth (averaging 3 percent per year).

Although there clearly is a widespread awareness that the energy problem is real and a growing determination to come to grips with it, there is only a dim awareness of the ways in which this problem affects the largest issue of all: the prospects for democracy, as it is experienced in the United States and as it is in question in the world at large.

Democracy has both a substantive and a formal meaning: it is a set of values and a belief in a process of self-government. The energy problem poses difficulties in both respects—difficulties that seriously affect domestic and foreign policy.

HOME TRUTHS

The Value Implications

Any society that has persisted for many years, that has survived inner turmoil and exterior challenges, and especially one founded in the name of certain ideals, is more than a geopolitical entity, more than an economic unit, more even than the sum of the symbols, rituals, and lifestyles of the people who inhabit it. It is also a community bound by shared values. Such is assuredly the case of the United States, which was founded as an experiment in republicanism, with the aim of protecting and promoting the belief in the "rights of man" as these were understood by the partisans of enlightenment and progress in the eighteenth century. The first Americans were also, on the whole, religious, and their convictions directed them not only to be concerned with the afterlife, but also to work diligently for the improvement of the earthly condition, as stewards of a higher will. The belief in individual liberty was bound together with a belief in personal and civic virtue, grounded both in secular and in religious convictions.

No more than other peoples, however, have those who settled this continent lived up to their highest ideals consistently or in every respect. Even the Civil War did not resolve the problem of relations between the races, and the social consequences of slavery still remain to be fully overcome. Nevertheless, although black Americans have sometimes been understandably bitter in gauging the distance between American ideals and American reality, they have come to believe in these ideals no less ardently than other Americans. The same conviction is shared by members of the ethnic minorities, which, during their experience as immigrants, have suffered persecution at the hands of xenophobic "nativists" and have been denied the equality of treatment guaranteed in the Constitution. While American society is less a melting pot than a mosaic of different cultures, religious communities, and subgroups of many varieties, there has grown up among virtually all of them a strong and widely-shared feeling of belonging, an often indefinable attachment to a common culture—the signs of which are discernible in similarities of attitude and behavior even more than in declared belief.

Politically, the feeling of belonging is expressed in a deep and persistent commitment to the values of democracy. The reality of this political consensus often goes unrecognized because so many of the particulars are taken for granted, such as regular and free elections and the legal protection of the rights of individuals. It is only when the everyday reality of American political life—with all its tensions and difficulties—is compared to that of societies in which there are no free and regular elections, in which the rights of individuals are unprotected, that the consensus over fundamental values becomes apparent.[4]

In this consensus over fundamental values and its major variations, certain assumptions are implied with respect to the basic conditions of social life. It has always been believed, not only here but elsewhere as well, that this country is, more than most others, a land of opportunity, the land of "unlimited possibilities," as Goethe put it. This opportunity has often been understood in material or economic terms, as the possibility of bettering one's own condition and of looking forward to the steady improvement of the standard of living for one's children and descendants. These improvements are thought possible not only because the resources of the continent are available for exploitation, but because the laws and the social system provide the stability and encouragement to assure room for initiative, for risk taking, and for sheer hard work. Opportunity has also meant the chance to live one's life differently, with respect to religious beliefs and practices and cultural pursuits. This opportunity arises not only because of the sheer size of the country, but also because of its diversity, and because of its constitutional and ideological commitment to toleration and cultural pluralism.

Earlier generations of Americans were not particularly conscious of the need to preserve the natural environment. For the most part, this goal was either ignored in the rush to industrialize or neglected on the ground that it was not a major problem. In recent decades, concern for the environment has risen sharply, in response to perceived threats to health and safety, and also as a result of changing standards of values. Americans have come to understand, more than in the recent past, that material affluence is not the only measure of social and personal fulfillment and that the preservation of the natural environment is essential to physical well-being as well as to aesthetic and psychological well-being. They have also come to appreciate the frustrating differences between the ability of a highly technological civilization to accomplish great feats in mastering the physical and biological aspects of existence and its apparent inability to understand or deal with human and social pathologies.

In all these particulars, there is a close connection between the values Americans hold and the means available for achieving them. It is this connection that the energy problem affects at many points. In order to identify those points where the impact is the greatest, it is useful to single out certain critical issues, even though in reality they form part of a continuous cluster of concerns. Among these are the link between economic growth and social stability; the definition of liberty, mobility, and fulfillment; the concern with personal security and the threat of violence; the prospects for political pluralism; and the impact of the energy problem on the concern for health and safety.

Economic Growth and Social Stability. For over a century, Americans have grown accustomed to a steadily expanding economy. Despite cycles of boom and bust, they have come to expect continued economic growth and with it an ever-improving standard of living. Although there is much dispute over the question of whether relative direct income shares have changed at all since the turn of the century, it is generally agreed that absolute levels of prosperity have advanced for virtually all income groups. At the same time, actual inequality has diminished because of the growth of public services and transfer payments during the past two decades. These benefits have considerably improved the relative economic conditions of certain subgroups, notably the very poor and the elderly.[5] The expectation of continuing improvement may account, in part, for the decline in industrial strife since the Great Depression of the 1930s. Prosperity also provides government with a revenue surplus, which can be used to cushion the effects of economic dislocations or (as in the case of the Apollo Project) to support large-scale ventures in national prestige building. To the extent that the energy problem will compel a slowdown in the rate of economic growth, it could affect not only the actual experience of improvement, but just as important, the expectation of continuing improvement. On both counts, the effect on social stability could be serious, even traumatic. "Relative deprivation" and the inability to achieve expectations can be even more potent sources of frustration and antisocial behavior than a more desperate level of poverty, which may only promote feelings of hopelessness and enervation.

In a "zero-sum society," as Lester Thurow has described the present prospect,[6] competition for the distribution of the social product could well become more intense. Instead of cooperating to increase the size of the economic "pie," the various economic interests would be even more anxious than they normally are to protect their own shares. Each group would aim to shift any necessary sacrifices onto the shoulders of others, and they would all have a reduced incentive to cooperate for the sake of a general improvement.

The general willingness to accept inequalities would be tested just as seriously in a long period of no economic growth as it is in periods of depression when there is an economic decline. Those least able to adjust, because of their economic marginality, would be tested most severely. As in depressions, those deprived not only of minimal means for economic security, but even of hope for improvement, could become resentful and violently angry. Even the frustration of a comparatively small number of talented individuals can have disproportionate social consequences, because they can become instigators of movements of social protest, or members of small but effective revolutionary or terrorist cells. Economic dissatisfaction is, of course, by no means the only motive for such activity. Some studies suggest that extreme social protest may be more congenial to the politically and psychologically "dispossessed" children of the middle and upper classes than to the young of lower economic classes.[7] Nevertheless, there is a significant correlation between impoverishment, especially in ghetto conditions, and crime and social deviance. Prolonged economic deterioration could make matters worse. Social order might require methods of repression only rarely experienced in this country, and the quality of urban life might deteriorate into one of even greater insecurity than now prevails. From Shays' Rebellion to the bonus march and "Hoovervilles" of the Great Depression and the burnings of inner city areas in the 1960s, there is ample and continuing evidence of the fragility of American civil order under conditions of extreme economic stress. The possibility that a relatively permanent "under class" may have developed, as a result of a combination of the shift toward skilled employment and the pathologies of modern social life, makes the prospect even more ominous.

Alternatively, American society might respond to a period of no economic growth by adopting radical measures of income redistribution aimed at forestalling serious deterioration in social harmony and social order. For this to happen, however, many more Americans than have so far responded to campaigns for such radical redistribution would have to be prepared to accept the sacrifices that it would require. The recent surge in tax revolts on the part of middle-income groups suggests that the initial response to deterioration is likely to be self-interested rather than altruistic, or at any rate that those who are used to fending for themselves will insist that if there is to be austerity, it must involve curtailment of public services and concentration on the job of restoring economic growth as a matter of the highest priority.

To avoid either the extreme of social breakdown or that of a defensive and highly repressive effort to maintain existing inequalities in the face of general economic malaise, it will obviously be necessary to maintain a positive rate of economic growth and to assure that minimal guarantees of economic security remain in place. The higher the rate of growth, the more it can be anticipated that friction and conflict over scarcity will be kept in check—though the problems of affluence will emerge to replace them, as they have already begun to do for certain segments of the population. To the extent that the availability of affordable energy is a factor in economic growth, it is a critical lever of social stability.

Liberty, Mobility, and Fulfillment. Directly and indirectly, the traditional availability of abundant and relatively inexpensive energy has enabled Americans to enjoy the "blessings of liberty" in a great variety of forms. Conversely, the restriction of energy supplies could affect the experience of liberty in various ways. The most cherished fundamental forms of liberty need in no way be affected. The rights to think for ourselves, to speak out without fear, to worship as we choose, and to govern ourselves do not depend on the availability of energy. In other respects, however, the constriction of customary options, including those that eliminate drudgery in the workplace and the household and allow ease of movement from place to place, could well be experienced as a loss of liberty. To a considerable extent, improvements in living standards have been a function of the availability of energy. A deterioration in these standards, many of which are now taken for granted, would be felt as a deprivation of liberty, insofar as liberty has come to be understood as the ability to take advantage of the many opportunities for satisfaction and fulfillment that a complex industrial society brings within general reach.

Two special problems have sometimes been raised in this connection, but neither of them is the most pertinent focus for a discussion of the links between energy and liberty. One is the possibility that civil liberties might have to be gravely curtailed in order to achieve security against sabotage, terrorism, and the diversion of dangerous materials from nuclear power plants. In Chapter 4, we noted that this does not seem to pose an especially acute problem. The other concerns the possible effects of rationing in energy emergency. In the absence of the social solidarity that war tends to generate, rationing might not be readily accepted as a necessary constraint upon individual liberty. Even in wartime, rationing systems spawn efforts of evasion. In the absence of war, efforts to get around restrictions might well be more common and carry less of a stigma. The imposition of centralized, bureaucratic systems of planning would surely be felt as a restriction of liberty, even if it did succeed in promoting optimal distribution. But rationing schemes have inherent disadvantages, which make it likely that market-oriented schemes would be adopted instead.[8]

The consequence for personal liberty that could be most serious is the possible effect of energy scarcity on physical mobility. In contrast to cultures that are centered on stable communities, and in which there is only a limited degree of movement from place to place, highly industrialized societies are characterized both by regular displacements of population and by far more movement in, around, and between places of settlement and occupation. Like all industrial societies, this country exhibits a very high degree of physical mobility in the ordinary conduct of life. Mobility can be experienced in the simplest way as the ability to live apart from one's place of work and to enjoy convenient transportation from home to work. In a more complex way, it is experienced as the ability to travel to distant places for business or recreation or to change jobs and way of life. Mobility in either sense is not necessarily the paradigmatic form of liberty, but it is certainly one of those forms of liberty that Americans prize, as is evident from the extraordinary extent of automobile ownership.

As automobile ownership came within the reach of almost every one, residential patterns expanded outward from the central cities to the suburbs. Systems of urban mass transit often fell into disuse and sometimes into bankruptcy as ridership declined and more and more people came to prefer, and to be able to afford, the convenience and flexibility of the private automobile. Encouraged by the construction of housing and roads, many people used the new mobility that the automobile allowed to move further away from their places of work. In some cases, the changing locus of work led people to move to the suburbs, as newer industries sought the horizontal space that was more readily available in the outlying areas, and as they became less dependent on the infrastructure and access to other businesses that inner-city locations provide.

Although "urbanization" increased in the period after the Second World War, what was really happening was that a considerable shift of residences and businesses was taking place from the central city to its satellite suburbs. The "metropolitan area" was an increasingly real entity, measured by the degree of personal and spatial interdependence and interaction, but the movement of activity described as urbanization was actually suburbanization. Densities declined steadily. Residential patterns became highly dispersed as housing tracts and subdivisions sprang up in what had once been rural small towns or potato fields. Downtown business centers were abandoned in the rush to find space in suburban shopping malls. Expressways built to link farms to market and region to region became crowded with commuters from dormitory suburbs to central city employments and attractions.

The explosion into the suburbs did not produce a corresponding change in political structure, for the most part. Politically, the metropolitan area remained fragmented. In the New York area, one study found no fewer than 1,400 governmental units.[9] Efforts to create metropolitan systems of government, incorporating a layered approach applying the principles of federalism and resembling those successfully adopted in Toronto and Winnipeg, were generally rejected. Families fleeing the central cities, and newly burdened with school construction and other costs in suburbs, refused to tax themselves to maintain or renew the central cities, and the states were unwilling to compel metropolitan political integration.

Because of the resulting pattern of dispersal, a return to widespread use of mass transit would not be economically viable. Even if the initial capital costs are subsidized, elaborate new systems like BART in the San Francisco area seem to have difficulty attracting enough ridership to support their operation or to warrant the high investment required even when other, indirect benefits are taken into account.[10] Mass transit remains less convenient than the use of private automobiles because suburban sprawl forces users of mass transit to make several connections and because the existence of road alternatives tempts those who keep cars for other purposes to use them for commuting. Inexpensive light-rail and bus systems are attractive alternatives, as are car pooling and the use of vans. A renewed emphasis on elaborate forms of mass transportation, however, is probably precluded, because of the absence of metropolitan planning and because of the unfavorable economic realities dictated by present circumstances. In these circumstances, the best hope for preserving opportunities for mobility in metropolitan areas seems to lie in a combination of incentives and disincentives, which would have the effect of shifting some of the burden to flexible and low-cost systems of mass transit, while encouraging the use of more fuel-efficient vehicles carrying more than one commuter.

It is at least arguable that restrictions on physical mobility may in some respects be a blessing in disguise, insofar as they promote a reexamination of the tendency to identify liberty with ease of movement, and especially with the use of the private automobile. The American "love affair with the automobile" may turn out to have been only a passing infatuation. In some respects, certainly, the "romance of the road" has already lost much of its allure. Modern freeways are more efficient but they lack the charm of winding country roads. The teen-age habit of cruising downtown has become as obsolete as downtown itself. Vacation travel in cars and in mobile homes remains attractive for many people, especially in view of the costs of alternative means of transportation and accommodations, but in other respects it may well be that the penchant for "hitting the road" reflects impulses that are the residue of an earlier era of frequent migration and that may be gratified just as well closer to home. We can only speculate whether the constriction of the cultural propensity for mobility, especially as it is made possible by the private automobile, would be accepted as a necessary adjustment, or whether it would be perceived as an unacceptable constraint on personal liberty. In this respect, as in regard to social stability, the most certain way to avoid the danger that frustrated expectations will lead to antisocial behavior is to maintain a sufficiently high rate of economic growth in order to keep open a wide menu of options.

Personal Security and the Threat of Violence. Like most societies, the United States exhibits elements of both civility and violence. Americans can boast of one of the longest records of the peaceful transition of authority in modern history. The prevalence of the rule of law and of a willingness to accept the obligations of citizenship is evident in such matters as the general compliance with tax laws and in the general respect for the civil liberties of unpopular dissenters. There have always been countervailing tendencies, however, and these have become more evident in recent years. An "underground economy" has developed in which the tax system is effectively bypassed. Concern over conspiratorial activities, terrorism, and violent crime has led to abuses of civil liberties by government agencies, a resurgence of vigilante groups like the Ku Klux Klan, and a widespread resort to efforts of self-protection. While there are a few direct links between the energy problem and the upsurge in violence, there could well be some association of an indirect nature, which is worth considering, especially on the assumption that the energy problem could become even worse.

Foreigners have been especially sensitive to the evidence of violence in this society. Because of our tendency to romanticize the past, we do need to be reminded that this country was born in an act of violent revolution. Although it is true that ours was less sanguinary than others, and much less a civil war than comparable revolutions in Europe, it nevertheless gave armed self-defense an aura of patriotic legitimacy that still survives, even though it is grossly anachronistic in the nuclear age. Frontier conditions only reenforced that aura and made violence seem a part of everyday life. The absence of settled conditions at the western frontier produced the legendary "wild west," with its sidearms and shotguns and its "great equalizer," the Colt .45. Even in the major cities, violence reached chronic proportions as masses of immigrants poured into overcrowded slums and individual gangsters and criminal syndicates preyed upon them and more settled communities. In recent years, as poor blacks and Latin Americans have migrated to urban areas, the same pattern has been repeated, except that this time drug addiction has intensified the crime problem, and the availability of weapons and the willingness to resort to force have both increased.

If the energy problem worsens, these trends may well be reenforced. Although there is no well-defined and demonstrable link between poverty and proneness to violent crime, the evidence of recent years suggests that conditions such as exist today in the central cities tend to be correlated with high rates of certain types of crimes, including personal assault. The ugly behavior experienced in such areas during power blackouts is an indication of what could happen more generally as a result of energy shortages. The violence that has erupted on gas station lines—in suburbs as well as central city areas—and at road intersections blocked by protesting truckers during the gas shortages is another indication of what might be expected from future shortages.

The indirect effects of economic difficulties produced, at least in part, by a failure to manage the energy problem, could be at least as serious. Already, the shortage of heating oil has produced hardship in certain regions and among the very poor. Hard times, when they are not so overwhelming as to unite a community, can instead lead to a breakdown of sympathy in favor of a more intense pursuit of self-interest. Whether civility will be maintained where it exists and restored where it has been lost will depend in part on how well this problem is managed. Intermittent energy crises will test the commitment to civility over violence. Acute shortages, especially of resources necessary to life, such as heating oil, are bound to promote regional self-interest, as is evident in the imposition of state "severance taxes" upon exported minerals and in resistance to coal mining and the construction of power-generating plants intended to supply distant users. Efforts to compensate producer regions will alleviate this tendency, but major efforts to satisfy energy needs will require a renewed sense of national interdependence.

Political Pluralism and the Role of Experts. A distinctive feature of American society is its "pluralism," a term that has several meanings but that in political usage denotes a society in which there is not only a significant variety of subgroupings (ethnic, religious, economic, regional, and cultural) but in which power is deliberately disaggregated rather than concentrated.[11] Pluralism in this political sense reflects constitutional principles and social structure. Among the major expressions of political pluralism is the system of federalism through which political authority is shared among the central government, the states, and the localities. Another is the separation of the public and the private sectors, and with it, the encouragement of privately owned businesses, churches, trade unions, philanthropies, and other voluntary associations, civic, athletic, political, and charitable. It is this devolution of power and initiative among a welter of agencies, some public, some private, some supported from centrally collected revenues, others local in scope and support, that is often considered to be the single most important guarantee of political liberty.

American pluralism has been under pressure in recent times from two contradictory tendencies, both resulting from the growing complexity of industrial society. On the one side, this complexity has spawned giant industrial concerns, often conglomerate and multinational, and other large-scale organizations, such as international trade unions. On the other side, it has generated "big government," as the political system has responded to demands for the regulation of enterprise in the public interest and for the provision of more and more common services, due to the enormous expansion of the role of military security and of welfare functions.

In conjunction with these organizational changes, both of which have promoted bureaucratization and a remoteness of power from those affected by it, has come another effect of advanced industrialization: the growing reliance on expertise. This too has increased the distance between the layman and the policy maker, in private as well as in public organization. The tendency toward bigness, toward anonymity, toward a sense of powerlessness on the part of the solitary citizen is further reenforced by the technology of mass communications. The role of the mass media puts the electoral center of gravity on national campaigns and removes it from the localities where political power was formerly dispersed. It also tends to erode the role of political parties in favor of electronic imagery and propaganda. Pluralism survives, but it is certainly under assault from these various tendencies. What happens as a result of the energy problem, and more specifically, what is or is not done to alleviate the problem, will have considerable impact on the prospects for a continuation of American political pluralism and the system of large-scale popular self-government that it helps to make practical.

General failures of social institutions promote a general distrust of all authorities, public as well as private, including those whose authority rests on the credibility of their claims to expertise. In a democracy, some measure of distrust of authority is healthy and invigorating, but carried to the extreme it can lead to a severe weakening of the bonds of social obligation. When these bonds break, acts of civil disobedience become more common, to an extent destructive of civil order, of habits of compromise, and of willingness to resolve social tensions. The storm of protest directed at the governmental and utility authorities in the aftermath of the Three Mile Island accident is a sample of what could become a political pandemic. In this case, it was still possible to relieve much of the understandable outrage by the appointment of a commission of inquiry and by the adoption of stiffer measures to prevent a recurrence. More such accidents, and above all a catastrophic accident, could affect not only the public willingness to accept the energy policies of governments and the utilities, but the willingness to accept authoritative pronouncements and the good faith of authorities in other respects as well. It is therefore not enough for energy policy to be formulated in the light of expert judgment of risks and benefits; policy making must also consider both the impact of catastrophic accidents on public confidence and the no less serious danger that failure to provide reliable energy systems could create critical shortages, which might also undermine public authority.

In order to reenforce pluralism, it has sometimes been suggested that the best energy strategy is one that emphasizes small-scale and decentralized technologies rather than those that require heavy capitalization and that lend themselves to large-scale construction projects requiring prodigious efforts of professionalization—and a comparable effort of government regulation. This argument has been raised along with others in support of solar technologies. One problem with this suggestion is that, for the time being, the most practical forms of solar energy (apart from hydroelectric power and biomass conversion) are those that provide water and space heating. It remains to be seen whether solar systems for generating electricity or for producing liquid fuels can be made sufficiently economical to make a substantial contribution to energy services in the near term.

Assuming that some of these systems do prove to be practical, they might well be more amenable to individual and community ownership and control than other forms of energy. Even so, it is likely that the same tendencies toward economic concentration that have made themselves felt in other areas of high-technology industry would also make themselves felt in the solar industry. At most we can anticipate that the private power utilities would have a diminished role in the provision of energy if solar energy takes a significant share of the market. Other cautions are also in order. Some utilities are cooperative in ownership; these presumably encourage pluralism. In all cases, utilities are subject to regulation by state utility commissions; as such they are a part of the system of federalism that functions to realize pluralism. A dispersal of energy sources to communities, moreover, could make the gap between affluent suburbs and central city ghettos even wider: the most affluent communities will be in a position to take care of their own needs and to resist efforts to tax them to pay for the provision of energy to the ghetto dwellers. As it is, utilities are under public constraint on a larger geographical scale and are required not only to provide services to all users but to spread the costs equitably.

In general, the importance of solar energy for pluralism ought not to be exaggerated. It is quite true that if it can substitute for nuclear energy and for fossil fuels as a source of electricity, there will be less need for great central power stations, and political and economic power may be more dispersed. It takes a very optimistic attitude to suppose that this will be possible on a large scale before the turn of the century. And in any case, the "solarization of America" will not affect the other tendencies toward centralization, which can be expected to grow stronger unless there is a general effort to unravel the complexity of industrialization, which sustains our high standard of living. If, in conjunction with the advent of solar energy, there is also a general return to small communities and lower expectations, the cause of decentralization would indeed be served—but this would require much more than the use of solar energy.

Health and Safety. With growing technological sophistication and growing affluence have come rising expectations with respect to public health and the provision of medical services. Overall, hazards to health and safety are lower than they have ever been, as is evident in falling mortality rates and rising average standards of health. New technologies, however, sometimes pose new and unexpected threats. In an effort to cope with such dangers, strenuous efforts have been made to protect health and safety, including the creation of such agencies of government as the Environmental Protection Agency, the Occupational Safety and Health Administration, and the Congressional Office of Technology Assessment. These efforts follow upon the earlier precedent of the establishment of the Food and Drug Administration and other government agencies charged with establishing standards in particular industries. The effort to formulate national energy policy raises problems of health and safety at virtually every turn, but these problems are sometimes exaggerated because the hazards from energy systems tend to be treated in isolation. Compared to the hazards from other normal activities, the sum of energy-related hazards is still small. Another reason the problems are of more public concern than other hazards is that expert opinion is divided or uncertain about their nature and seriousness.

The most controversial aspect of this problem has been the issue posed by nuclear energy. Nuclear energy raises in the first place the problems associated with emissions of low-level radiation from mining, processing, power production, and the disposal of wastes. But as the Ford-Mitre study points out, these risks must be compared with those from coal-fired plants, so long as these plants remain "the principal alternative for electric power generation for the rest of this century."[12] The same study points out that the degree of ionizing radiation (the most damaging form) to which the general public is exposed from all activities connected with nuclear power is, on the average, only a tiny fraction of normal background radiation and other man-made sources, such as radiation from diagnostic xrays. In 1970, the average dose rate in millrems per year was only .003 from nuclear power, out of a total exposure of 210, including 130 from natural sources and 72 from diagnostic xrays. It may well be that radiation from radon in airtight buildings poses a more serious threat than nuclear power. It does not follow, however, that nuclear energy should be regarded as benign simply because there are other more serious sources of hazardous radiation. Compared to other forms of electricity generation, nuclear energy is safer if by safe we mean the probability of related accidents. Coal mining and the burning of coal account for more injuries and fatalities than nuclear installations. These installations, however, are disadvantageous insofar as they pose the (low probability) risk of high catastrophe in the event of a major accident. And there are still great uncertainties about the long-term health hazards from radiation exposure at the levels experienced by workers at nuclear facilities (who are now allowed exposures of 5 rems per year) or by those exposed to accidental releases.

Other alternatives pose other hazards. Liquid natural gas terminals located at or near densely populated areas are also potential sources of catastrophic danger. If coal burning increases on a global scale, irreversible damage might be caused by the release of carbon dioxide to the atmosphere, causing climate changes that could have incalculable effects. A reliance on oil, on the other hand, means a reliance on tankers and offshore rigs, with the resulting increased likelihood of the pollution of the seas and the destruction of beaches and wildlife. A general reliance on fossil fuels, whether by mining surface coal or by exploring for oil and gas on public lands previously reserved as wilderness or recreation areas, poses other threats of environmental degradation. Still, the storage of nuclear wastes poses unique health hazards, because certain components in these wastes remain toxic for thousands of years. No matter how well they are secreted, there cannot be perfect assurance that they will remain permanently contained. With careful containment, however, the likely risk to health (through possible contamination of ground water) would appear to be very low and of limited consequence. (See the discussion in Chapter 4.)

The political problem of resolving health and safety issues may, paradoxically, be greater than the technical problems. Without a long experience of exposure to radiation from nuclear power generation, for example, it is hard to obtain a degree of assurance of its comparative safety sufficient to justify a commitment to rely much more heavily on nuclear power. Similarly, although technical experts have devised a variety of workable systems for the storage of nuclear wastes, there is no way they can demonstrate with certainty that these systems will assure safe containment for thousands of years, no matter what happens to the earth's geology or—much more uncertain still—to its social and political systems. The only way to reach consensus is therefore to assess *probable* risks and benefits *comparatively,* bearing in mind the uncertainties that must enter into their assessments. There is no way to achieve perfect certainty in making these assessments. The most that can be expected is that the major alternatives will be carefully explored and the decisions made by accountable authorities.

Energy Policy and the Democratic Process in Theory

The energy problem poses an unusually serious and ramified challenge to the democratic process, if only because it requires widespread public appreciation and a coordinated approach to a diffuse set of issues on which there is bound to be considerable disagreement and conflict of interest. Even in the best circumstances, democracy rarely works in practice as it is designed to as an ideal. Under pressure, and especially in cases like this, where there is no single definable crisis situation, the democratic process is tested even more seriously than it is otherwise.

Under normal circumstances, democratic systems are not simply devices whereby the majority rules and the rights of individuals and minorities are protected. Since many people do not choose to exercise the suffrage, electoral majorities often turn out to be pluralities formed by electoral minorities. Participation rates are apt to be lower in local elections than in more general elections. The result is that elections to office and referenda on issues can be considered democratic only by a definition that allows for relatively low rates of participation. Most policy questions, moreover, are left to be resolved by an indirect or representative expression of the popular will. In practice, modern democracy is a system in which actual decisions are made by elected and appointed officials who are expected to be accountable to those who choose to vote.

Even this modified view needs to be further qualified to take account of the reality of large-scale democratic politics as a process of group organization and interaction. Policy outcomes are usually greatly influenced, and sometimes determined, by the interplay of highly interested groups, even more than by electoral pluralities, which give only a general direction to the actual process of policy formation. It is with good reason that political scientists often prefer to think of democracy as a system of pluralistic coalition building.

Policy is also influenced by those who achieve power by virtue of the offices they hold, both elective and appointive. Civil servants and judges are expected to be guided by constitutional and statutory law, but their independence of electoral accountability leaves them considerable room for the exercise of independent judgment in determining what these laws actually mean in particular cases. Regulatory agencies exercise "quasi-judicial" authority, sometimes in accordance with legislative mandate, sometimes in ways that reflect the independent initiative of the agencies. The same is true for the role of the judiciary, which extends its initiative by virtue of the claim that other agencies default in exercising their responsibilities.

Much policy is said to be made by de facto combinations of organized interests, legislative alliances, and administrative agencies. The result is sometimes described as the working of an "iron triangle" in which reenforcing pressure is brought to bear from all three points.[13] Thus, agricultural policy is said to be the resultant of the interaction of organized groups of farm interests with their supporters in Congress and with their administrative defenders in the Department of Agriculture and elsewhere. Defense policy is similarly said to be influenced (though in this case by no means determined, because other interests and agencies are significantly affected) by a similar triangulation among defense contractors, the armed services and their research and procurement installations, and their legislative supporters on appropriations and military affairs committees.

American democracy is not only representative and pluralistic, but also federal, based as it is on a system of coordinate sovereignty in which the states exercise powers constitutionally reserved to them, except where an overriding national interest leads to preemptive actions by the central government. Even though modern conditions (notably the expansion of interstate commerce and the impact of foreign policy) have led to more and more such preemption, there are still many areas where state initiatives remain important. There is no federal system for incorporating businesses, for example, or for regulating the insurance industry. In both cases, business interests have been successful in preventing federal preemption. In energy policy, the states exercise a variety of powers, including the power to tax resources exported to other states, which affect national policy.

The energy problem is a particularly challenging one for this system of government because, in the first place, it is a ramified problem, which lends itself particularly well to fragmented rather than coherent treatment. It is, in this sense, tailor-made for interest-group politics and for a corresponding fragmentation of effort in the executive and legislative branches of government. Policy must be shaped both in terms of aggregate goals, such as the ratio of total energy usage to the rate of economic growth, and in terms of a host of particular issues, such as the balance of risks and benefits in connection with each source of fuel. Account must also be taken of foreign policy considerations. A degree of long-range planning is required from politicians, government agencies, and public and private enterprises, which are all more inclined to concentrate on the short term.

In certain respects, moreover, the energy problem is attended by so much uncertainty that it poses a classic case of decision making in conditions of uncertainty. Examples abound in virtually every direction. We have a reasonably good idea of how much coal is available within our own borders. We are less certain about the future costs of extracting this coal and about the health and environmental effects of increased reliance on coal. The projected cost and availability of oil depend critically on the effective demand for it at various price levels, and the price levels are in turn influenced by the production levels set by the supplier countries as well as by the rate of discovery and exploitation of new sources. Policy for nuclear energy depends on the resolution of such still controversial issues as reactor safety and waste disposal. The future demand for energy services will depend upon the degree to which shortages and price constraints will affect consumption, by promoting conservation and greater efficiency in usage. Perhaps the greatest uncertainty of all surrounds the many new technologies under development, from solar generation of electricity to the extraction of oil from shale and tar sands to fusion and the production of liquified hydrogen by any of a number of possible techniques. Which of them will prove economical and by what dates is so uncertain that none can be safely factored into future supply—yet neither can they be neglected in the expectation that they will all be of no short-run consequence. Nor can we be certain of all their environmental impacts in advance. In view of such uncertainties, projecting future supply and demand is bound to be hazardous in the extreme.

Even if all these elements of uncertainty can be kept within reasonable bounds, there remains the problem of defining the social goals that are to guide policy. If these goals are to reflect a desire to satisfy prevailing expectations, including the expectation of continuing improvements in standards of living, it will certainly be necessary for this country to pursue policies aimed at achieving high rates of economic growth. Such a goal presupposes high rates of energy consumption—though there is good reason to expect that moderately high rates of growth (of the order of 3 percent per year in real terms) can be achieved with a ratio of energy services to output lower than is now the case. This can be accomplished by substituting other factors of production for energy and by achieving greater industrial conservation and efficiency. In addition, a willingness of consumers to accept modifications in lifestyle—smaller, more energy-efficient homes, "downsized," lighter, more fuel-efficient cars, and energy-saving household appliances—could also help significantly to allow the economy to grow without increasing consumption of energy.

Although the development of a general goal is crucial to the formulation of policy, it is not one that can easily be made "holistically" or by referendum: it will be registered incrementally if at all and it will reflect market constraints, government policy decisions in particular areas, as well as individual expressions of choice. Without some public commitment to a general goal, however, the necessary long-term planning cannot be undertaken to phase in new systems and to invest in expansion and modification of existing facilities. Although radical commitments in conditions of uncertainty can prove to be very mistaken and expensive, a failure to anticipate needs early enough and to initiate projects that will come to fruition only later can also be very costly.

If it is important to have some degree of consensus over goals, it is also important to appreciate the effectiveness of the various alternative means of implementing policy. Depletion allowances and the decontrol of prices, for example, are expected to stimulate exploration, whereas a tax on "windfall" profits and other measures may have exactly the opposite effect. Some policies may be adopted that would encourage or mandate a shift to the use of coal, while others, by dampening the increase in price of other fuels or requiring expensive "scrubbing" to remove pollutants, may have a countervailing effect. Permitting utilities to "pass through" fuel costs to consumers through the "fuel adjustment" provision encourages necessary investment, but limitations on their ability to pass through capital costs act to discourage replacement of oil-fired generating capacity by coal or nuclear facilities, even when the replacement would be economically justifiable in terms of total long-term costs to consumers. Federal loan guarantees and cooperative investment in research help stimulate the development of synthetic fuels, but strict enforcement of antitrust laws may serve to inhibit development. The debate over the means to be used will not be eliminated by agreement on a general goal, but at least it will provide a clearer standard by which the effectiveness of policy options may be judged.

Energy Policy and the Democratic Process in Practice

As yet, there cannot be said to have been a successful effort to frame and implement a coherent national energy policy. Under presidents Nixon and Carter, brave attempts were made to declare such policies, but in neither case was public or legislative support extensive enough to move to draft and institutionalize such a policy. The Carter effort came close to actual implementation when it led to the adoption by Congress of a package of measures, though not the entire comprehensive plan presented by the administration. Certain of these measures remain intact, but the grand design has been significantly weakened already by the actions of the Reagan administration. The general aim of the new administration is to rely on increasing supply by stimulating market forces rather than on subsidizing synthetic fuel developments or mandating conservation—a considerable departure from the premises of the Carter policies.

Especially in view of this about-face, it remains largely true that American energy policy has not been framed comprehensively but by sector. The actual evidence indicates that it is much easier to propose a comprehensive policy than it is to bring together the coalition of interests necessary to establish it. The conflict of interests—sectoral, regional, and economic—is one obstacle, and the controversy among experts is another. Until recently, national energy policy can be said to have been the unintended result of sectoral policies. There have been five such sectors: oil, coal, nuclear energy, natural gas, and hydroelectric power. Each of them has had its cluster—or triangle—of interested parties, represented by lobbyists, government agencies, and legislative friends. The result was that efforts were made to formulate "fuels policy" in each sector but no comprehensive energy policy. "It is not clear," Craufurd D. Goodwin concludes from a historical study of U.S. energy policy, "even with the benefit of hindsight, just what the nation should have done to prepare for the energy transitions that lay ahead. What is striking, however, is that with very few exceptions persons taking part in the discussion seldom had any broad national interest at heart." Moreover, he adds, new or unconventional modes of energy supply, "which were by definition without any corps of supporters and defenders, were bound to lose out to entrenched interests." Cost comparisons inevitably favored established technologies since, in the early stages of development, new technologies almost always require heavy investment and produce comparatively expensive products.[14]

Given this pattern of sectoral isolation, it seemed only logical to create an Atomic Energy Commission rather than a more comprehensive department of energy. In retrospect, this decision has been criticized for generating an imbalanced approach to the development of energy policy within the federal government. Even though the AEC was also supposed to investigate other new technologies, including solar energy, the fact that its primary mission was the development of nuclear power predisposed it to ignore other possibilities, and alternatives, among them solar energy, were left to the agencies to support. The custom of dealing with energy sector-by-sector resulted in the commitment of very substantial public subsidies to petroleum, coal, atomic energy, and hydroelectric power, while other possible sources of energy supply were given only meager encouragement. The more recent decisions to create the Energy Research and Development Agency and then the Department of Energy obviously reflect the view that a more comprehensive strategy is preferable. But even if the department should be dissolved, the rationale for a comprehensive policy would remain persuasive, especially in view of actual experience.

In the absence of a comprehensive approach, energy policy has often been internally inconsistent and anomalous. The decision to exempt intrastate natural gas from federal price regulation resulted in a two-tier price structure that distorted market forces and led to less than optimal patterns of use. The low price of natural gas led to less than economically rational use of this premium fuel. Shifts to coal have been encouraged and even mandated without corresponding efforts to assure that coal demand would be developed and that transportation would be available. After pursuing a policy of "drain America first," the federal government reversed course in the 1960s and increased dependence on imported oil. Utility regulation, by structuring profitability in terms of return on investment, has given the utilities an incentive to make large capital investments, and perhaps too little incentive to be concerned with escalating costs. Now that environmentalist pressures have made certain of these investments highly problematic, the regulatory agencies are making it difficult for the utilities to attract the capital they need to make critically needed investments.

The process by which these decisions have been reached is one that is very much affected by the interaction of interested groups. It is no exaggeration to say that energy policy has been decided by the interplay of powerful forces in each sector, without an adequate effort to impose any overriding sense of the national interest. This is not to suggest that the producers and distributors of the various forms of energy have been the only organized interests with an impact on policy. Mine safety regulation is the result of pressure by trade unions. Environmental regulation is the result of pressure from environmentalists. Each of the special agencies in the Department of Interior sought to define and protect the interests of the various sectors they represented. In general, however, it is this sort of public and private group interaction rather than some clearly-defined effort to ascertain the broader public interest, that has generally determined energy policy. As Goodwin observes,

> The American experiences with energy policy over the past forty years... must give one cause for concern about the capacity of government to deal intelligently and effectively with such a series of challenges. What we witness above all is a Congress dominated by relatively narrow special interests, both regional and industrial, an executive branch riven by bureaucratic conflicts of all kinds, and a presidency distracted constantly by "larger" issues, and with a time horizon normally extending not beyond the next election and often much less.[15]

In an effort to minimize the role of interested forces, a campaign has been mounted in recent years to increase public participation in the making of energy policy, especially in connection with nuclear power. Those responsible for the campaign have argued, plausibly, that decisions to proceed with such installations have often been taken by default rather than after active and careful public scrutiny. They contend that democracy requires that those most affected by decisions should be more directly involved in the making of the decisions. In the case of nuclear installations, they contend, people living in the area of the facilities have not been consulted in advance or properly informed of the risks. The experience of accidents, especially the accident at the Three Mile Island reactor, and the publicity given to alarms over the dangers from radioactive emissions and potential malfunctions of the emergency cooling systems, have aroused support for efforts to require consultation with local residents.

In general, state and federal licensing procedures are designed to provide opportunities for public participation. The utilities and their supporters argue that the review process is being misused as a device to postpone constructive actions of any kind. It is certainly true that when the legal opportunities are exhausted, opponents of nuclear power have not always been inhibited from taking direct action to block construction, as in the case of the demonstrations at the Seabrook, New Hampshire facility.

The antinuclear movement has raised questions that go beyond the issue of nuclear power and its environmental impact. What is in question is the nature of decision making in democracy. There are technical questions in dispute, such as the significance of low-level radiation hazards, the likelihood of accidents, and the adequacy of radiation-control standards. But there is another class of questions that politicizes the issue. These include "problems of human fallibility: the difficulties of waste disposal and transportation; the diversion of plutonium for military purposes; the possibilities of espionage and terrorism, and the consequent need for surveillance...." Using these as a basis, antinuclear forces have raised larger issues: "What kind of society is implied by a nuclear economy? What are possible costs of a nuclear program to future generations? What level of risk is society willing to accept? How can an equitable distribution of risks and benefits be guaranteed? Does the government have the institutional capacity to manage long-term risks? Ultimately, who should be making such decisions?"[16]

At first, as Dorothy Nelkin and Susan Fallows point out, the debate was largely over technicalities, and the antinuclear groups were fragmented. The passing of the National Environmental Policy Act (NEPA) and the implementation of the act in the *Calvert Cliffs* decision, which required the AEC to "extend its responsibility to include environmental concerns," gave greater legitimacy to the protests, and assured the protesters that courts would consider their views. Court decisions also liberalized the rules of standing to allow suits to be pressed by individuals without economic injury, to be advocates of the public interest.[17]

The effort moved from challenges of the AEC to forced publication of data and opening of hearings, and from particular campaigns to a general political effort. Thus, there were eight state referenda beginning with the California instance in 1976. All eight failed. Although all these efforts fell short of their largest goals, they succeeded in publicizing the antinuclear case, in promoting the passage of stiffer state laws, and in stimulating public awareness of the controversial character of nuclear power. Further efforts followed, including the campaign by Common Cause, which complained of the close links between industry and the AEC and found that 71.5 percent of the 429 senior staff of the NRC had been employed at one time by private industry in the energy field and that 90 percent of these came from companies with NRC licenses, contracts, or permits.[18]

The government agencies, under congressional prodding, have taken steps to broaden participation opportunities, but Nelkin and Fallows conclude that because of restrictions on participation and initiative, "Intervention becomes essentially token, for people can only express reactions to policies already made."[19] There is legislative opportunity for greater participation, by advisory committees and intervenors, but the opportunities are restricted, and efforts to provide intervenors with funds have failed. Lack of financial resources is a serious handicap. These efforts to expand participation have been criticized because they build in costs and delays and cause proceedings to be inundated with irrelevant information; they invite everyone to pose as an advocate of "the public interest," diffuse responsibility from agencies to public, and discourage innovation.[20]

By a political definition, effective management of the energy problem must include a reasonably successful effort to make use of the institutions of democracy in achieving and implementing policy decisions. Decisions will be required that will not always be popular with consumers and with significant interest groups. Consensus over the validity of the process by which decisions are reached is therefore crucial to their general acceptance. Several considerations are important in maintaining this consensus.

The legislative review of energy policy should be vigorous and critical. The Joint Committee on Atomic Energy was accused, with some justification, of being the "Siamese twin" of the Atomic Energy Commission rather than its legislative watchdog. The critical or adversary role of the legislature vis-à-vis the administration offers assurance that public interests are being protected.[21] Only when this function is performed well can public confidence in the integrity of the process be maintained. The problem is that members of the specialized congressional oversight committees tend to become advocates of the activities they oversee. In the process of becoming particularly well-informed about a particular technology, they are often prey to a subtle form of co-optation, which leads them to defend the technology against critics. In addition, such committees tend to attract legislators whose districts contain major facilities designed to develop or exploit the technology.

Safety reviews should be undertaken by independent agencies. The safety reviews conducted by the old Atomic Energy Commission violated this principle. The results were always open to the charge that they were designed to serve the agency's interest in promoting atomic power—a charge that had earlier been made with some justice in the debate over fallout from weapons testing. The creation of an independent Nuclear Regulatory Commission amounts to a recognition that an agency charged with the promotion of a technology is not always the best one to entrust with the judgment of its safety. The same principle might well be applied in other areas of energy policy.

Public officials, especially those charged with regulating and evaluating the impact of energy facilities, should be as forthcoming as possible both about potential benefits and risks. In practice, this is a more difficult demand upon them than it might seem to be. Efforts to "bend over backwards" to be completely candid make it easy for critics to take particular statements out of context. Nevertheless, an effort at studious impartiality on their part is essential, if public confidence in governmental reviews is to remain high. Otherwise, official statements will be greeted with cynicism, and citizens will be tempted to give equal if not greater credence to less well-founded judgments. The problem is that once the agency takes a decision to approve a design or grant an operating license, it becomes committed to the project and its public statements are apt to reflect a defensive rather than a strictly objective position. If difficulties appear later on, those responsible for the initial appraisal will be tempted to become even more defensive.

There is unfortunately no way to escape the dilemma posed by the fact that in most cases regulators must be drawn from the industry they are asked to regulate. Regulation of safety, health, and environmental risks requires considerable expertise in regard to the industry being regulated. This requires not only formal training but actual on-the-job experience, which can only be acquired during employment in the industry. The training provided in engineering programs at universities is aimed to prepare students for work in industry, and the faculty who provide this training must consult with industry in order to retain their familiarity with the "state of the art." Only if the morale of the government agencies is strong and if the regulators are imbued with a strong sense of professional and civic responsibility can the inevitable influence of an industrial background be counteracted. It might also be helpful if the engineering faculties were drawn into regulatory work as regularly as they were into industrial activities; some of the resulting "socialization" to regulatory problems might lead to a more balanced education of engineers. At the same time, it could also result in a more rigorous regulatory system.

Politicians and publicists also need to make an effort to explain as clearly as possible the issues in controversy. It is not easy to explain the possible effects of windfall profits taxes or price deregulation, especially since well-informed experts disagree about them, but the effort to broach these complexities to the public is essential in order to allay suspicions of conspiracy and to invite response. Ordinarily, it is at least tacitly understood that in a representative democracy highly technical issues must be left to be decided by elected and appointed officials, who will be expected to consider expert opinion and who will be held accountable for the results of their decisions by the electorate. Referenda among ill-informed voters are hardly likely to produce better results, any more than acts of civil disobedience, however well intended. If the aim is to maintain a workable form of democracy, the best approach is one that emphasizes representative responsibility in the context of vigorous public information and debate.

Contrary to the gloomy forecasts of the prophets of technocracy, the increasingly technical character of many of the most salient social questions has not made democracy obsolete. Most of these questions involve judgments about policy goals as well as about technicalities. Often the technical issues are uncertain, and there is disagreement among the experts. In these conditions, it would be foolish to rely only on the judgment of a single technical advisor, as C.P. Snow rightly warned.[22] The adversary process, which is at the heart of the parliamentary system of government, is a far better system for resolving policy disputes, even those that are highly technical, provided that the process is not distorted and overwhelmed by conflict among interested groups and provided some effort is made to assure that all qualified opinions are well presented.[23] Democracy, however, imposes heavy demands upon citizens and officials. In a technologically complicated age, those demands become greater than ever.

In order to meet these demands more effectively, it has sometimes been suggested that new modes of conflict resolution need to be introduced. One model is provided by the experience with the regulation of recombinant DNA research. The scientists who understood the possible dangers acted responsibly to alert their colleagues to them and to propose safeguards, which were adopted by a government agency. National and local reviews were instituted involving experts and ordinary citizens. Although new controversies have emerged over unregulated industrial activities in the burgeoning field of biotechnology, the issues surrounding university research were resolved with reasonable success: the scientists were able to continue working, and restrictions were adopted until safety could be more certainly assured. Equally important, those who might be affected by the risks had the opportunity to become informed of the issues and contribute to the shaping of policy.[24]

Another experiment has been undertaken in an area involving energy policy. The general intention is to replace the adversary process with an alternative also derived from legal practice, which is referred to as the "rule of reason," and which attempts to bring the parties to a controversy together in the hope that they can resolve their disagreements by discussion and negotiation in order to avoid costly and protracted litigation. The National Coal Policy Project sponsored by Georgetown University is an attempt to test the possibility of substituting a rule-of-reason approach for the resort to litigation, which has become endemic in controversies over environmental impact. Representatives of industry and environmental groups have been brought together in two different "caucuses" under the guidance of a professional mediator. In 1978, after a year of work, conference task forces reported to the plenary group, which adopted a series of consensual resolutions that demonstrated a remarkable degree of agreement among normally antagonistic groups. Perhaps predictably, however, the results did not satisfy all industrial or environmental groups, especially those not directly involved in the proceedings. Critics argued that such important public issues should not be resolved in small-scale, private forums but in the ordinary institutions of society: courts, legislatures, and administrative hearings. They also objected that certain groups with high stakes in the outcome (including labor unions and Indian tribes) were not represented.[25]

Further experiments such as these may well be helpful in improving the capacity of democratic systems of government to deal with highly technical questions, where necessary by upgrading capacities for policy research and analysis. Many steps have been taken in recent years in this direction. In the area of energy policy, the departments of the federal executive have enlisted expert advice from many sources, including the major study produced by the National Academy of Sciences–National Academy of Engineering and resulting in the CONAES report. The Congressional Office of Technology Assessment has also undertaken a large number of studies of particular issues. The institutions of government in this country are now well-equipped to secure qualified advice. What is done with all the studies depends on the intelligence of the officials who must make use of them and on the support they can generate within the voting public.

9 America and the World

VALUES, INTERESTS, AND THE LIMITATIONS OF POWER

In domestic terms, conflicts of values and interests occur within a framework of a single social and political system. Ties of commonality and interdependence can be relied upon to moderate these conflicts. It is the absence of such a framework that makes conflict less amenable to coordinated resolution on a world scale.

Studies like the elaborate analysis undertaken by the International Institute of Applied Systems Analysis (IIASA)[1] show plainly that the world as a whole faces acute imbalances between growing population and a variety of essential resources. These include not only fossil fuels but also other resources necessary for nutrition, shelter, and economic growth. The same studies show, however, that the gravity of the problems varies with region, resource endowment, and level of affluence. To succeed in coordinating efforts to deal with the resource problems of the world as a single entity, the nations of the world will have to set aside or overcome formidable differences of interest as well as of ideology. Although they share the same planet, and therefore must all ultimately rely on the same finite set of resources and the same ecosphere, these nations remain competitors and adherents of rival political blocs. This political and economic reality cannot be ignored.

For the United States, the global energy problem must impinge on domestic policies because our interests and value concerns are far flung, and because we are a major economic and military power. American policy is therefore bound to have significant effects upon people everywhere. At a minimum, our policies must take account of the impact of our behavior upon our allies, our adversaries, and those who cannot be considered either allies or adversaries but whose cooperation we wish to assure.

In a politically volatile world rife with nuclear weapons, the prevention of wars, especially wars that threaten superpower confrontation, must be the foremost concern of foreign policy. The energy problem poses the danger of war in various ways. The gravest danger is probably the escalation of conflicts over access to energy sources, especially petroleum. There is also the possibility that the spread of nuclear power could increase the danger already posed by the proliferation of nuclear weapons.

National interest requires access to supplies of energy adequate to permit the defense of vital interests and of a standard of living necessary at the very least to assure social stability. In some respects, these national interests of the United States coincide with those of our allies; in others they come into conflict with them. The adjustment of competing interests must be a constant preoccupation of American foreign policy. This country and its allies also have a keen interest in achieving better relations with the noncommunist developing countries, by helping them to achieve economic growth and social improvement by means that avoid the repressiveness of the totalitarian model of development.

Often, considerations of interest shade imperceptibly into considerations of values. A secure and prosperous alliance of free nations is one in which the prospects for stable and vigorous democracy are enhanced. Sometimes, however, interests and values come into conflict. Nuclear war can be prevented only by accepting the need for compromise with adversaries committed to ideologies and practices we may find disagreeable and even abhorrent. Alliances may need to be maintained with regimes that do not adhere to the standards we ourselves profess. But value considerations cannot and should not be entirely ignored or overridden: we are bound to be drawn to the side of those nations that share our convictions and to be cool or hostile to those that do not. The energy problem, as a global issue, compels us to think about the extent to which our values should guide our policies. Is there a democratic ideal that should guide us, or should such a concern be dismissed as either naive and utopian or dangerously unrealistic?

Even assuming that energy policy could be based on some synthesis of interests and values, it would still be necessary to recognize that in the world as it is presently constituted, no nation, no matter how powerful, can act entirely as it might wish to do. Even great powers must appreciate the limitations of their ability to influence and control the behavior of others. The energy problem poses the problem of the limitation of national power acutely. Having become dependent on foreign suppliers of oil, we and our allies have become vulnerable to the use of the "oil weapon" as an instrument of political pressure and to the regulation of supply as a means of compelling a transfer of economic resources from the user nations to the suppliers. To this extent, our political and economic fortunes have become hostage to the will or mercies of others. The vastly superior military power of the consuming states could easily be brought to bear to break this stranglehold, but only at serious and uncertain risk.[2] Even a "surgical" military operation could lead to spreading and uncontrollable conflict. The threat of military action could lead to acts of sabotage in the oil fields or the ocean straits through which oil tankers must pass, which might cripple extraction and transportation facilities and lead to prolonged shortages. The possible need for a long-term occupation of the oil fields would revive a form of outright imperialism, on the model of the *Pax Romana* and the British Empire, which would not only be difficult to maintain but would represent a grave setback to the cause of self-determination and international cooperation, which is critical to the success of the cause of democracy in the world.

American power is no less limited in relations with allies. Unless they are to be treated as satellites or colonies, in which case they would have little incentive to remain loyal allies, they must be accorded the same right we claim for ourselves to determine their own national interests and to pursue them even when they do not coincide with our own. A democratic alliance must be one in which mutual respect prevails, and in which all partners accept the principle that persuasion and give-and-take rather than coercion or blackmail must be the basis of the alliance.

With respect to our adversaries, all use of power cannot be foresworn, but it must be carefully calibrated in accordance with a hierarchy of objectives. If we were to renounce all resort to the instruments of power, we would have only ourselves to blame if our interests were put in jeopardy wherever they might lie. To defend those interests, however, we cannot rely on resort to physical force alone, not only because that force might be wasted on objectives of unequal importance, but also because whatever we stand to gain may be far outweighed by the consequences of war. A willingness to use means short of physical force, such as embargoes on strategic goods or the use of other economic levers, is therefore essential. The principle of deterrence requires that an adversary be restrained by the fear that physical force will be brought to bear, not that the force itself actually be used. When an adversary chooses to commit a hostile act against a less than vital interest, we must be prepared to defend that interest with whatever means are appropriate—that is to say, with a limited and sometimes indirect use of force.

It is sometimes argued that the best way to defuse the threat of East-West conflict is to promote economic interdependence. The problem with this thesis lies in understanding the difference between interdependence and a lopsided relationship that only increases vulnerability to aggression by means other than invasion. An arrangement whereby the Western powers become suppliers of capital and high technology to the Soviet bloc in exchange for Russian natural gas, and little else, is not well calculated to promote a sense of *mutual* dependence. Once in place, the pipeline will be entirely in the control of the Soviet Union, and the Western powers will have a heavy financial stake in its operation, as well as a heavy dependency on the fuel it will provide. The Soviet Union will no longer need either the technological assistance or the capital it now requires. It is equally true, however, that a lessening of economic incentives for cooperation could increase East-West tensions and induce greater reliance on costly measures of military preparedness on both sides.

What general principles, then, should guide the United States in developing foreign policies to address the energy problem as an international issue? One alternative would be to put aside national interests and values as parochial and accept the definition of the energy problem developed in such studies as that of IIASA. In these studies the energy problem is defined in terms of the general world need to manage global resources. The aim is understood to be that of maintaining general world economic growth while achieving a transition from reliance on fossil fuels to more sustainable sources of energy, notably nuclear and solar sources. If such a perspective were to be adopted, the United States would have to commit itself to policies that take just as much account of the needs of the developing countries and the Eastern European countries as of domestic needs. Instead of pursuing much greater independence of foreign suppliers, we would have to enter into cooperative arrangements whereby all world resources would be pooled and distributed in accordance with priority needs to be established by international agreement. All questions of self-interest and ideology would have to be subordinated to a concern for the global interest.

In some respects, such an approach is appealing, not only on grounds of idealism, but even on grounds of pragmatic self-interest. The more the United States would compel itself, or be compelled by international agreements, to cut its reliance on imported oil completely, the greater would be its political freedom of action and the better would be its relations with the developing countries. They would have greater access to the oil they need, their balance of payments problems would become manageable, and the U.S. economy would become strong enough to import the goods they would be able to produce.

Global cooperation would also make it possible to monitor and control environmental risks associated with the energy problem. Unless there is a coordinated effort, for example, to monitor the effects on climate of CO_2 emissions from the burning of fossil fuels, it is possible that efforts to maintain high rates of growth could cause major climatic changes, the effects of which are unpredictable. The IIASA estimate is a possible increase of between one and four degrees centigrade. In this respect, the disjunction between the world system of separate sovereignties and the unitary ecology of the world's atmosphere is dramatic and disturbing: "We are faced with the dichotomy of having a highly disaggregated political power on the globe and the truly global problem of atmosphere CO_2 buildup. Are we doomed to encounter this dilemma? Probably, yes." The IIASA analysts make this point in the hope of stimulating awareness of the danger. The CO_2 problem is certainly one issue that poses such an overriding danger that all nations ought to cooperate, at the very least in carefully monitoring the CO_2 buildup.

In other respects, however, it is arguable that an effort to develop energy policy based on the assumption of world interdependence is unrealistic, in view of the actual disaggregation of power and the various types of conflict that divide the peoples of the world, and would not necessarily promote the values to which we are attached or the national interests on which the realization of those values depends.

Our interests and values, however, will not be properly defended if we pursue a parochial policy that takes no account of the needs of others or the limitations of our own capacity to influence events. The best approach is to define the international energy problem in such a way as to identify our aims as a member of the community of nations, but with a special responsibility for the welfare of our own citizens and our allies and for the cause of equal liberty in the world. What this set of principles means in practice can be clarified by examining it in terms of our relations with allies, with the developing countries, and with adversaries.

ENERGY AND THE ALLIANCE OF FREE NATIONS

Because the major allies of the United States, especially Japan but a majority of the Western European countries as well, are even more dependent than we are on imports of fossil fuels, cleavages have developed between them and ourselves with respect to energy policy. The Europeans, especially the French, have committed themselves to nuclear energy and have rejected American concerns about the impact of international sales on nuclear weapons proliferation. They point out, understandably, that the United States has sold reactors to nonsigners of the nuclear nonproliferation treaty (Egypt, Israel, India, and Spain), and—less understandably—they have not scrupled at providing reprocessing and enrichment facilities, even though these facilities may be used to obtain fissile materials for weapons. France and Germany have agreed to supply such facilities because they are anxious to make their own nuclear industries commercially viable by breaking into "an otherwise unassailable American market."[3] In response to American urging, however, our European allies have drawn back from certain of these efforts, notably in the case of Pakistan. The Pakistanis have nevertheless been able to secure much of the technology they need from European sources. In addition, the European nations have sought to make their own separate arrangements with Middle Eastern oil suppliers and have been reluctant to commit themselves too strongly to alliances of consuming nations, for fear of offending their suppliers. They have also tailored their policy on the Middle East (in particular the Arab-Israel conflict) to placate the Arab oil-producing states, even though it has meant weakening the unity of the alliance on this score.

In general, the divergence of energy policy has shown that although our allies depend upon the United States as the military and economic bulwark of the alliance, they neither feel they can depend upon us to develp energy policies that will take due account of their interests nor wish to subordinate their independence to the alliance in all respects. They are both allies and economic competitors, and they are clearly anxious to maintain a measure of independence of the United States, even if it means creating strains in the alliance. As a result, the only energy agreements so far entered into by America and its allies are relatively weak and uncontrolling. The oil agreement covers only emergencies and otherwise leaves each nation to its own devices. As a result, the supplier nations do not confront an equally united group of consumers. Prices are kept high and storage capacities are strained. The agreement to control nuclear proliferation is too weak to be effective and barely provides the signatories with a face-saving document to show that there has been consultation on the issue. Recriminations among the allies follow from the lack of harmony in energy policies. The United States criticizes the French for selling nuclear reactors to unstable regimes like Pakistan and unfriendly regimes like that of Iraq. The French counter by criticizing the United States for failing to adopt more strenuous measures to conserve oil and for thereby putting the rest of the world, including its allies, at continued high risk. The asymmetry of energy dependency made the Europeans and Japanese reluctant to join in stern reprisals against Iran when the American diplomats were taken hostage by fanatics in Teheran in 1979.

In general, the Europeans tend to argue that American energy policy aims to maintain the competitive economic advantage for its manufactured goods that this country has traditionally enjoyed because of the availability of relatively inexpensive energy. American exports of goods produced with petrochemicals are a particularly inviting target for such complaints. Controlled natural gas prices in the United States, it is argued, amount to an export subsidy to the U.S. chemical industry. The countries of the European Economic Community have therefore accused the United States of unfairly "dumping" chemicals and synthetic fibers in European markets. Similar accusations have been raised in connection with U.S. aluminum exports to Japan.

This intense economic competition among America and its allies has badly undermined efforts to curb the proliferation of nuclear weapons. The Nuclear Nonproliferation Treaty has proved to be only a limited and porous barrier. American willingness to sell "sensitive" facilities without requiring stringent safeguards against diversion of fissionable materials for the development of weapons set a bad precedent, which other supplier states were quick to follow. When, under the Ford and Carter administrations, alarm developed over the prospects for proliferation, which were being engendered by efforts to sell nuclear power systems, America's European allies balked at our efforts to make such sales more restrictive. The Europeans suspected that American efforts to discourage the sale of reprocessing and enrichment facilities and the development of breeder reactors were really motivated by our concern to prevent them from acquiring a competitive economic advantage over our nuclear industry. They did agree, however, to accept a tacit moratorium on new sales during the International Nuclear Fuel Cycle Evaluation. After two and one-half years of study, however, this effort did not produce a "technical fix" that could prevent the plutonium designed to be used in power reactors from also serving as material for weapons. In any case, it now appears that a policy aimed solely at controlling the sale of power facilities (with or without reprocessing and enrichment capacities) cannot prevent states with sufficient resources and determination from acquiring nuclear weapons. A policy aimed at regulating and controlling the fuel cycle can still be of some benefit in curbing nuclear proliferation, but in general, as Smith and Rathjens suggest, "the best hope of stemming nuclear proliferation lies in dealing effectively with the motives that lead nations to want to have nuclear weapons."[4] American security guarantees and collective agreements establishing nuclear-free zones would serve the purpose better than attempts to control the spread of nuclear capability. As Smith and Rathjens also suggest, an American energy policy designed to reduce the oil dependency of our European allies would also help, by making them less anxious to sell nuclear technology to oil-producing states.[5]

The same economic disunity has arisen in connection with East-West relations. Efforts to rally the NATO nations to join in applying severe sanctions against the Soviet Union in the event of a Russian invasion of Poland had to contend with the keen interest of the Western European countries not only in detente in central Europe and in trade relations with the East bloc countries, but also in obtaining supplies of Soviet natural gas. The investment that would be required to build the pipeline to obtain these supplies would further inhibit the Western European countries from taking stands that would jeopardize the investment and continued access to the gas. Already such vast loans have been made to the East bloc nations that the Western economies have developed a vested interest in the political stability of the communist regimes in these countries, despite the Western interest in their liberalization. East-West energy deals might serve the cause of peace by promoting interdependence, but they would also weaken the ability of the allies to resist Soviet moves elsewhere in the world or to support dissenters like the Polish trade unionists who seek to change the repressive social and political structures of the countries of the Soviet bloc. To provide an alternative, however, the United States will have to take the interests of its allies into account more seriously in formulating energy policies, perhaps by promoting reliance on American coal instead of on Soviet natural gas. As of now, this option is certainly not practical, but if it were considered of sufficiently high priority, it might be made more feasible by investments in American and European ports, rail facilities, and coal-burning plants (and synthetic fuels facilities) instead of in a Soviet pipeline.

The United States faces a no less delicate problem in adjusting its energy policies to the interests of its two closest neighbors, Canada and Mexico. Since both have significant resources of oil and gas, both are vital sources of supply for this country. Each has economic and energy needs of its own, however, to which these resources are critical—which are not always compatible with American interests. The United States has tried to entice both countries into a continental energy arrangement, but neither has been willing to consider such an integration.

Since the United States is the closest and biggest customer, however, both countries have an interest in selling surplus supplies even in the absence of some grand plan of coordination. The Mexicans are reluctant to agree to increased sales of energy lest it cause runaway inflation, due to difficulty in absorbing the proceeds fast enough in development projects. If the United States would collaborate more actively in accelerating the pace of Mexican development, it is at least conceivable that this problem might be alleviated. The Canadians have a somewhat different concern. They are reluctant to deplete energy resources that may well be needed to sustain future industrial growth in Canada. They are particularly concerned that the Canadian economy is too heavily based on extractive industries and has not yet developed enough industries to process and export as finished goods the raw materials with which the country is so well endowed. They are also anxious to increase the extent of Canadian ownership over these valuable resources. But further exploration and the development of secondary, manufacturing industries can occur only if there is sufficient capital, entrepreneurial risk taking, and markets for the products. A collaborative approach could make it possible for Canadians, in effect, to trade some of their energy reserves for a more balanced economic strategy. The achievement of this sort of cooperative arrangement will require creative initiatives on both sides. The present climate tends to promote protectionism, which is not in the best interests of the parties, with the exception of such ad hoc arrangements as the trans-Canada pipeline. This pipeline, it should be noted, was agreed upon only after the Alaska pipeline was built first, a pipeline that did not bring oil to the markets in which it was needed most but was built first because of "security" considerations. In this important instance, Canadian nationalism stimulated American nationalism, to the detriment of both countries.

RELATIONS WITH DEVELOPING COUNTRIES

Because the rise in energy prices affects the developing countries even more than advanced countries, and because the economic difficulties caused for the advanced countries by these price rises affect their ability to aid and trade with the developing countries, the gap between the rich and the poor nations is apt to be widened as a result of the energy problem. The new power of the OPEC countries vis-à-vis the advanced nations has made them champions to people in many of the developing countries. Some have been willing to support the oil producers and follow their lead, in the largely illusory hope that they might be able to imitate the precedent of OPEC by forming resource cartels in other commodities. Even failing this, they hope that a general reorganization of the world economy will take place as a result of the pressure of OPEC in which a "new international economic order" will be created. Such a new order, its advocates maintain, would end the unfair advantages that the rich have over the poor in market transactions by interposing a political authority in which all nations, regardless of wealth, would enjoy the benefit of equal sovereign authority.[6] This leverage would be used to compel a revision of the terms of trade and an equalization of opportunity. Raw material and commodity prices would be indexed to the price of manufactured goods. Access would be guaranteed to markets in the developed countries for the manufactured exports of developing countries, where these enjoy a comparative advantage. Tariffs discriminating against raw materials processed in developing countries would be removed. Foreign aid from the developed countries would be gradually increased, and proprietary technologies would be more available on favorable terms. These and other proposals would be designed to overcome advantages now enjoyed by the advanced industrial economies over those of developing countries.

There are grave internal contradictions in the case for a new international economic order. The campaign aims to create a more equitable distribution of wealth among nations, but it is silent about the need for more equal distribution of wealth within the developing countries. Nor would it allow for outside interference in internal affairs, even to assure protection of human rights. Although its advocates claim that its goal is an equalization of opportunity, there is reason to suppose that if a genuine equality of opportunity failed to produce satisfactory results, political pressures would seek to promote redistributions not justified by contribution to economic welfare.

These deficiencies make it more likely that the United States and the other advanced countries will attempt to persuade the developing countries to imitate the example of Taiwan, South Korea, and Japan by adopting Western-style modes of development. Those who become relatively successful would be "co-opted" into the Western economic system, as Saudi Arabia has been for all practical purposes. It may be that the alliance between the OPEC nations and the others in the United Nations "Group of 77" will erode because of internal strains, as the failure of the OPEC nations to alleviate the plight of the developing countries, caused by higher oil prices, begins to become more manifest. The OPEC nations could act to forestall this breakup, by undertaking a serious program aimed at redistributing some of the wealth they are presumably gaining as "reparations" for previous exploitation on the part of the other formerly colonial countries. If this is done, and if OPEC acts as a kind of development clearing house, it is conceivable that the developing nations could take significant strides toward achieving some of the goals of the New International Economic Order (NIEO). Even so, this could not be done without the cooperation of the advanced countries, as suppliers of technical assistance and as customers. As a practical matter, it must be regarded as an unlikely option.

As things stand, the condition of the developing nations threatens to grow significantly worse in many cases. The advanced countries can choose to avert their gaze and to rationalize their indifference by invoking the principles of "lifeboat ethics," "triage," etc. In some cases, such a policy will run no great risks. In others, it will produce such instability as to threaten the peace of the world and the interests of the United States. If Haiti's economy becomes even more incapable of providing for its own needs than it has been, Haitians will wish to emigrate in even greater numbers. Control of such immigration to this country would require very controversial measures. The same is true for illegal immigration from other Caribbean and Latin American countries.

In some cases, economic instability poses a threat of revolution and of Cuban-style communist takeovers, in Latin America and elsewhere. The United States and its allies therefore will be pursuing a short-sighted policy if they try to ignore the plight of the developing nations. To do otherwise, however, will require adopting energy policies that take due account of the needs of these countries. This may well mean foregoing certain energy supplies vital to the needs of the developing countries, or making a much more serious effort to assist them to use all appropriate technologies to solve their problems, including financial and technical assistance in oil exploration. Such a positive approach could have vast dividends, by promoting peaceful cooperation on terms acceptable to American values, and by laying the groundwork for a much more effective system of world cooperation than has so far been achieved. But this option will require sacrifice and a sophisticated public understanding of the issues.

ENERGY WARS

If the energy problem grows even worse, major wars could be fought over access to scarce supplies, or to overthrow cartels like OPEC, if their pricing policies deprive the industrial nations of access to vitally needed energy supplies. Energy supplies may well be considered so essential to survival as to warrant war, under the right of self-defense. Indeed, it has already been argued that the United States and its allies could justify military occupation of the Persian Gulf states on precisely this ground.[7] The proposed Rapid Deployment Force is designed to diminish some of the danger. Ostensibly, the RDF would come to the aid of Middle Eastern states in the event of threats posed to them directly by the Soviet Union or by its proxies. The line between providing assistance to a legitimate government and intervening in the internal affairs of a country we are protecting may be difficult to draw in practice, however. Nor are the governments of these states apt to be easily persuaded that the forces to be deployed in the area would under no circumstances be used against them. The danger of a major war over energy supplies is therefore not so remote as to be totally discounted.

It is more likely, however, that wars will arise to which energy is a contributing factor. For one thing, the availability of energy resources means that countries unaccustomed to great wealth suddenly have it thrust upon them. In their desire to behave as wealthy nations do, they are tempted to arm themselves with the latest weaponry and to display their power by resolving long standing disputes with their neighbors by using some of their new weaponry. The behavior of Libya in Chad and of Iraq in the dispute with Iran over the Shatt-al-Arab waterway illustrates the point. Indeed, the wealth that energy resources have brought to Iran has been at least an indirect contributing factor to the rise of revolutionary Islamic militancy in that country, as a backlash to the modernization that the Shah was attempting to bring about on the basis of oil wealth.

The random distribution of energy resources results in an almost automatic sense of separatism and conflicting interest. "God gave the earth to men in common," John Locke observed, "but he meant it for the use of the industrious and the rational." Whatever the intentions of the creator, however, the resources of the earth have been divided without obvious regard to Puritan virtue, thanks to the creation of separate sovereign nations in the course of history. There is little that can be said to be sacrosanct about political borders. They have changed considerably even in the present century. Nevertheless, attempts to develop international cooperation must proceed on the basis of respect for the political status quo, at least as a formal set of divisions. In energy policy especially, nations tend to stress their right to their "patrimony," especially if their natural endowment is a generous one.

Those less well endowed are apt to be more sympathetic to the view that the resources of the earth belong to mankind and therefore should be shared equally regardless of where they happen to be discovered.

The rights and wrongs of such arguments are difficult to establish, especially since territorial claims are often the result of migration or conquest and depend, at best, on claims derived from long settlement and the investment of capital and labor, and since the energy resources in question may have been discovered and developed by foreigners. Practical considerations often dictate that well-established, long-standing, and generally recognized claims should continue to be respected, especially since territorial claims often acquire moral qualities. Nations are more than territorial entities: they are moral communities, groupings in which rights and duties are established according to certain principles of legitimacy. This is a more difficult criterion to apply than the practical test, which boils down to a finding that some one person or group holds effective power over a given area. It is open to challenge when claims are made that authority does not in fact rest on legitimate foundations (as in a revolution or civil war or secession). Nevertheless, it is a necessary consideration if policy is to be made in accordance with moral as well as practical considerations. It may be practically prudent to come to terms with a regime that cannot be considered morally legitimate, but it cannot be a moral obligation to come to the rescue of a regime that is considered illegitimate—oppressive, imposed, dedicated to inhumane policies. Moral criteria are very hard to apply consistently, in view of the difficulties of deciding between competing claims and in view of conflicts between principle and national interest. Nevertheless, these considerations cannot be completely neglected without violating fundamental commitments. American energy policy must take account of the national interest in securing access to oil supplies for the United States and its allies. At the same time, the national interest must be defined to include concern for the values of democracy elsewhere in the world. In the long run, our security depends on ties of belief as much as it does on ties of interest.

The energy problem has also compelled us to recognize that there is a grim paradox in present international relations. The scarcity of energy has intensified claims of national independence, at the same time as it has made everyone more acutely aware of the need to recognize global interdependence. The oil-rich nations have experienced a surge of national power and pride and have attempted to use their wealth to increase their power and prestige. Nations that were quiescent colonial outposts for centuries have become supremely conscious of their rights and powers. The availability of energy has even reenforced separatist claims, in the case of Scotland and Quebec, making subnational autonomy a realistic rather than merely a romantic possibility. Even when separatism is not in question, regionalism is, as in the tensions in Canada between Alberta and the rest of Canada.

These centripetal tendencies are being experienced, however, at a time when the pursuit of narrow, parochial self-interest by all the nations of the world puts all of them at risk. Rarely has the principle of the tragedy of the commons been more poignantly illustrated. If enough oil-poor nations decide that they must burn coal, it is at least conceivable that the world as a whole (including the oil rich) will suffer catastrophic consequences. If the policies of OPEC should trigger a major depression in the Western advanced industrial countries, not only would their own petrodollar investments in these countries suffer, but even their ability to extract oil revenues would be gravely impaired. They must take care both to maximize revenue and preserve markets for their products. Conversely, an effort by the developed countries to undermine the stability of the governments that now make up most of OPEC could unleash such turmoil that the oil fields would be in jeopardy. They must balance their outrage at having to pay usurious prices with their concern to assure stability of supply. All of the relatively rich nations can choose to ignore the plight of the desperately poor, which is being made worse by increasing energy prices, but they do so at their own peril: these nations are sources of other raw materials they all need, and they are potential markets that will be needed if their economies are to continue to expand. They are also sources of international tension, which could drag outsiders into a world war, or at least into very costly local wars. Some form of international cooperation is therefore dictated even by considerations of self-interest. But these considerations are not the same for all the actors. Each of the various blocs has different interests and therefore a different strategy for playing the game.

It is easier to sketch scenarios of turmoil and tragedy than of international collaboration. In a century that has witnessed two major world wars and a host of sanguinary local wars, it would be too much to suppose that planetary peace and cooperation are just around the corner. If the consequences of these devastating wars have not taught men to turn swords into plowshares, how can we expect that to happen as a result of the shortage of energy—especially since the shortage does not affect everyone to the same degree or in the same way?

This very lack of "criticality" affects our approach to the energy problem as a nation, and it affects the world's approach to it even more. Whatever good reasons the systems analysts may present for modelling the energy problem as a world problem, and for dividing it into regions that take some account of political divisions, it does not follow that a rational division of labor will automatically take place, or even that the nations of the world can be convinced, by appeals to their self-interest alone, that such a division is advisable. It may well be more likely that they will fail to reach agreements, except insofar as these are plainly dictated by dire emergency and common need. It is worth speculating what the failure of any efforts at significant agreement could mean.

A continued steep rise in energy prices as a result of growing scarcity will certainly affect the prospects for economic growth, especially in the developing countries. In these countries, the poor often already live at or below marginal levels of subsistence. They need energy not only to heat their homes and power their vehicles but to be able to grow their food and to cook it. In Asia, as Roger Revelle has pointed out, three-fourths of the population relies on grains as the principal dietary staple. Grains cannot be produced in sufficient quantity without fertilizer, without power for irrigation, without fuel for agricultural machinery, without pesticides and herbicides and chemicals with which to treat the soil.[8] The grains cannot be consumed as food unless there is fuel to cook them. In India, Revelle notes, cooking and other household uses consume more than half of the total energy use.[9] Fuel from hydrocarbons is not the major source of the primary energy used in the rural areas of Asia, but it is a critical component in the overall balance of supply. The steep rises in oil prices, coupled with deteriorating demand in the industrial countries for the exports of the developing countries, have upset the precarious balance of energy supply in many of these countries. In the process, survival itself has been put in question for great numbers of the world's poorest inhabitants.

Because the conditions of most developing countries are so bad to begin with, the rise in oil prices has had effects that are little short of devastating. Some countries classified as developing, it is true, are actually benefiting, like the Persian Gulf states, Algeria, Venezuela, and Mexico, but there are 85 other countries in this classification in Latin America, Asia, and Africa that are net importers of oil. In almost 50 of these, 90 percent or more of commercial energy—derived from fossil fuels and hydroelectric sources rather than gathered from animals and forests—relies on the use of oil as a primary source, either directly or for generating electricity. In the others, oil dependency ranges between 50 and 90 percent.[10] These 85 countries include some that are developing successfully, like Brazil, South Korea, and Singapore, and others that have a much poorer record. Oil is a vital commodity for all of them because it is more flexible and versatile than other forms of primary energy and because it is a fuel that may be acquired without large investments in infrastructure and without the capital investments for exploration and development of other domestic sources. As the price of oil has risen, more of the export earnings of the developing countries have had to be used to pay for it. In Turkey, for example, imported energy cost 80 percent of all earnings on exported manufactures in 1977. As a result, an economy already stagnant was dealt a crippling blow, and unemployment went still higher than it already had been, up to as much as 20 percent. Brazil has had a record of growth admired by other developing countries, but Brazil is critically dependent on imported oil. In 1973, Brazil imported 800,000 barrels of oil a day. By 1979, oil imports rose to 960,000 barrels a day. But in the intervening years, Brazil's bill for imported oil went from $600 million to $7 billion, and as a result the rate of inflation soared and the rate of growth declined.[11]

Unlike many other developing countries, Brazil has had the ability to borrow and to reduce its oil dependency by producing alcohol as a substitute liquid fuel. All the cars in Sao Paulo now run on a mixture of 20 percent ethanol, 80 percent gasoline. By 1982, 15 to 20 percent of all cars built in Brazil were to be equipped to run on 100 percent ethanol. But Roger Revelle calculates that even though Brazil plans to devote 8 million hectares of arabic or potentially arable land to biomass to be converted into ethanol, this effort will provide the equivalent of a liter of gasoline per person for Brazil's estimated population of 150 million in the 1990s.[12] The ethanol will be a big help, but it will not solve the energy problem even in Brazil, which, unlike most other developing countries, can devote a large fraction of its land to biomass production without hurting food production. For the time being, Brazil remains dependent on imported oil.

For many of the LDCs, the single most serious consequence of the rise in oil prices has been the worsening of the deficit in food production. The Food and Agriculture Organization (FAO) predicts that except in Latin America, the developing countries will experience "a calamitous drop in food per capita" during the coming decades, unless present trends are reversed. The quality of food available to the world's poorest people, according to the FAO, will be inadequate to permit children to reach normal body weight and intelligence or to permit adults to engage in normal activity and enjoy good health. In Central Africa, food production is projected to be more than 20 percent below the minimum FAO standard, assuming no recurrence of drought. The World Bank estimates that the number of malnourished people in the developing countries could rise from 400–600 million in the mid-1970s to 1.3 billion in 2000.[13]

Another consequence of rising energy costs is that fuelwood is being depleted in certain areas where it is the major source of available energy. In the arid regions of Africa the gathering of fuelwood has become a full-time occupation. Where demand is concentrated in cities, surrounding areas have already become barren for large distances. The *Global 2000 Report* suggests that conditions may become even worse. "For the one-quarter of humankind that depends primarily on wood for fuel, the outlook is bleak. Needs for fuelwood will exceed available supplies by about 25 percent before the turn of the century."[14]

The governments of these countries face agonizing choices. If they pass on increases in fuel prices to consumers, for example, by raising the price of kerosene used for cooking, they may cause serious hardships and provoke riots and demonstrations, such as have occurred in the Philippines and Jamaica. If they do not pass on the increases, energy costs must be subsidized with capital critically needed to sustain development, and consumers will not be given economic incentives to conserve energy. An ever-tightening credit squeeze forces them to borrow more and more from foreign creditors to pay for oil imports and to service the mounting debt. In 1978, the oil-importing developing countries incurred combined current account deficits of $27.1 billion as a result of the additional cost of oil imports. In 1980, the deficit amounted to $68 billion. Annual debt service charges went from $10 billion in 1973 and 1974 to over $30 billion in 1978, and since then, they have undoubtedly risen even higher as oil prices have climbed again.[15] The total accumulated debt of the developing countries has risen from $50 billion in 1970 to $400 billion in 1980 and is expected to climb to $500 billion in the coming year.[16]

Balance of payments deficits as a result of payments for oil have been a problem for the industrialized nations as well, but here too the actual impact has been far more severe in the case of the LDCs. When the first oil shock was felt, there was fear that the impending transfer of vast sums of money to the oil-producing states would cause economic collapse in the industrial nations. What has actually happened is that most of the money paid out has been returned in exchange for goods and services or as investments. Great sums of "petrodollars" have been successfully "recycled" in this way in the wake first of the 1973–74 price increases and then of those of 1979. In the process, there has been a significant transfer of wealth from the industrial states to the oil-producing states, though the amount transferred is actually less in real terms than is sometimes supposed, because the inflation of prices the oil producers must pay for industrial goods and the deflation of the currencies they have had to accept in payment have eroded their gains to some extent. This is precisely why the Saudis have tried to persuade the other members of OPEC to agree to a stable set of formulas controlling production and tying prices to the rates of inflation in the industrial consumer countries.[17] The banking systems of the industrial states have been put to a severe test but they have survived it, and the economies of the industrial world have not been gravely undermined. As the oil producers acquire more of a stake in the economies of their principal customers—they already have $200 billion invested in Western economies, according to the *Wall Street Journal* (June 10, 1981)—they must take care not to cause major disturbances. Like Bedouin shepherds, they must keep their lambs shorn but not so naked as to be in danger of dying from exposure.

The condition of the oil-importing developing countries is far more precarious. They do not produce significant amounts of the petrochemical and other industrial goods, including advanced weapons systems, which the oil-producing states are interested in importing. Nor for the most part do they offer attractive opportunities for investments of petrodollars.

If large balance of payments deficits and debt service obligations are a problem for the United States, they are a potential catastrophe for the developing countries. Of the total developing country debt of $400 billion, about half is owed to commercial banks.[18] As these banks decide to refuse further credit, these countries can go only to the multilateral lending agencies, the World Bank, and the International Monetary Fund. These institutions of last resort are bound to become more and more reluctant, to require stringent conditions, and to run out of capital to lend. What will happen if there are significant defaults on their loans? According to one analyst, "Default by the poorer nations could precipitate a worldwide financial crisis similar to the U.S. banking crisis during the Great Depression. . . ."[19] Both the industrial states and the oil-exporting countries have every reason not to allow this to happen. If they listen to their financial analysts, they will establish an insurance fund to rescue threatened banks or take other steps to forestall a wave of developing country defaults. But the industrial states who must bear the greatest burden will not contribute enough loan capital to enable them to achieve significant rates of economic growth because that would permit OPEC to raise prices with impunity. If the world economy continues to be sluggish, private lenders will become even more reluctant to invest in the developing countries, which will become poorer risks than ever, as they become altogether immobilized by the vicious cycle of poverty.

The prospect is not a pleasant one, either for the developing countries or for those of us sufficiently farsighted enough to have chosen to be born in more affluent societies. The fact is that the economic interests of the industrial countries are bound up with the stability and the prospects of the developing countries. President Mitterrand of France, who understands this linkage well, has pointed out that 40 percent of the exports of the nations of the European community go to the developing countries. In our own case, the fastest growing market for our exports has also been in these countries. Over the 1970s, U.S. exports to them grew by 22 percent, compared with 15 percent for exports to industrialized countries.[20] If the developing countries could grow at a faster rate, they would be better able to buy more food and industrial goods from the industrial countries. It is therefore in our economic interest to take their needs into account in planning energy policies and international trade policies as well.

Both in the developing countries and the oil-dependent advanced countries, significant domestic economic problems resulting from energy problems could trigger hardship, civil disorder, and in some cases revolution and counterrevolution, and a general deterioration in the conditions of civilized life. What will follow from these disorders is too speculative to predict with any degree of confidence and will probably vary from place to place. On the basis of actual experience, it is plausible to expect an increase in selfishness and a corresponding decline in solicitude for the welfare of others. Unlike war, an energy panic does not necessarily inspire sacrifice and devotion to community against the outside threat. It is more likely to promote internal breakdown and anomie. It will produce general demoralization, as those who have come to expect or at least hope for improvement find their expectations dashed. It will produce aggressive behavior on the part of nations, which will come to feel justified because they act out of necessity. It will probably bring to power those who exhibit attitudes of toughness and concern for the self-interest of the nation above all.

It is worth recalling the terrible economic conditions and the sense of cultural paralysis that gave rise to the Nazi movement in Germany. No historical contrast could be more striking—or more instructive—than that between the moderate political style that prevails in both Germany and Japan as they enjoy unprecedented prosperity and the militaristic regimes that arose in both countries under conditions of economic adversity. In both cases, moreover, the rise of nationalistic and militaristic regimes was preceded by an extended period of intense economic protectionism in the industrialized nations. The Japanese attack on Pearl Harbor was precipitated by an oil embargo against Japan that its leaders judged to be a threat to its survival.

It is entirely possible that for various reasons related to the energy problem wars will break out that will have wide repercussions. Few predicted the war between Iran and Iraq, but in retrospect it is possible to understand it as a consequence of the new importance of energy resources and the failure to integrate these resources into political patterns of stable development. Iran under the Shah sought to use its oil resources to enable it both to develop internally and to achieve hegemony in the Persian Gulf. When the internal discontent unleashed by the Shah's politically illiberal style of development resulted in the fall of his regime, and the weakening of Iran as the policeman of the Persian Gulf, the way was open for another new "oil power," Iraq, to use its wealth in an effort to redress its grievances against Iran (resulting from the Shah's assertion of regional power) and to follow Iran's precedent by attempting to assert its own hegemony. The turmoil released by the reaction against modernization is another factor that may be anticipated to play a continuing role and that will not be easily placated even by economically successful development. (On the contrary, it seems to be correlated with great strides in secularization due to development.)

Other scenarios are conceivable, including a superpower war in the Middle East over access to scarce oil resources. Such a war might result from the powers being dragged into local conflicts, or it might develop out of conflicting ambitions, for example, a Soviet ambition to reach warm water ports and at least a share of the oil resources now committed to the West, by extending the arc of its influence from Afghanistan into Iran, thus creating a kind of pincer on the north using both Iran and Iraq. Even short of actually going to war against the NATO countries, the Soviet Union could achieve a position of strength from which to extract political concessions. The United States has already taken one step toward committing itself to such a conflict by declaring access to the oil of the Persian Gulf a vital interest of the United

States and its allies warranting military measures if necessary and by establishing an emergency force to make that threat of intervention credible.

Even in the absence of war and of infectious civil disorder, the failure to achieve some degree of international cooperation in the case of energy will be harmful in other respects. It will set back the efforts to achieve greater unity among the world's peoples and to recognize the fragility of our common biosphere, and the necessity to take common action to preserve it. It will set loose forces of competition and parochialism that will undoubtedly be felt in other respects, so that the already tenuous links between the peoples of the world will become further strained. Efforts to curb population growth by assisting development and spreading the latest public health standards, along with improvements in literacy, will all be jeopardized. Failures in these critical areas will further compound the problem, by creating political conditions in which efforts of reform and cooperation will become even harder to achieve.

These are grim possibilities. They need not come to pass, providing the world's leading industrialized democracy succeeds in putting its own house in order, by developing an effective energy policy, and providing reasonable initiatives are taken to encourage cooperation among those peoples of the world who share at least a minimum of our own commitment to democratic values.

IS THERE A DEMOCRATIC WAY TOWARD INTERNATIONAL COLLABORATION?

Because of its commitment to democracy, the United States should approach the energy problem not only as a matter of national self-interest but also as an opportunity for the extension of democratic principles to its dealings with others. This certainly means that we must be more aware than we have sometimes been of the need to respect the autonomy of other peoples.

The energy problem represents an opportunity as well as a serious problem, because it could make it possible for the United States to launch a major initiative on behalf of an effort both to alleviate world poverty and to promote democratic values. This will depend, in the first place, on the success of a domestic energy policy in relieving our own present gross dependence upon imported oil. By reducing this dependence, we will ease the hardships of others and promote a greater willingness on their part to cooperate with us. In particular, the European countries and Japan could probably be induced to enter into cooperative arrangements with the United States if we for our part pledged greater efforts to reduce consumption of oil. If we did so, we would be in a better position to restrain nuclear weapons proliferation, especially by strengthening safeguards to be enforced by the International Energy Agency and developing systems for the storage of plutonium. If we make a major effort to make our own abundant coal resources more readily available, not only by unclogging the present supply bottlenecks, but by actively expanding the extraction and shipment of coal, and by developing synthetic liquid fuels from coal, we could greatly alleviate the current world shortage of fossil fuel. Efforts to increase the domestic supplies of gas and oil could be equally important in promoting an improved international climate for cooperation. If we improve our bargaining leverage, we will also be in a better position to put pressure on the OPEC countries to redistribute more of their "windfall profits" to the developing countries.

Assuming that we can take these beginning steps, it may then be possible to move on to a more ambitious effort to create a global energy regime, or perhaps a global resource regime. It might be difficult to treat energy in isolation since those countries with energy surpluses are not likely to find that they have much to gain from becoming part of such a regime, unless it also takes into account their dependence on other resources. How ambitious the regime is will depend upon the willingness of the large nations to undertake such a cooperative effort. It could include advanced technology, and it may have to. The greatest concern of the developing states is certainly obtaining access to the advanced technology of the West. Precisely such access could be held out as an inducement for their collaboration in the creation of a resource regime.

It would be an illusion, however, to suppose that the direct creation of such an ambitious arrangement will be easy. The difficulties attending the law of the sea negotiations serve clearly to show the problems that lie ahead. Here, the effort was to create an ocean regime for the common exploitation and management of deep sea resources. It has been a kind of trial run for a more ambitious effort, and it may well not succeed. The difficulties stem in part from the differences of national interest at play—the differences between the landlocked states and those that depend on sea resources already and have come to regard them as available to the taker, the differences also between those nations that are in a position to exploit the resources and those that are not. The rich seagoing states have been unwilling to make their capital and technology hostage to the management of politically vulnerable transnational bodies. The poorer nations see the effort to protect private deep-sea mining as an effort to gain legitimacy for a "grab" of the last remaining "common" of the world. It is a classic modern standoff, which will be broken either by a compromise that respects the economic power of the rich nations and the political power of the poor or by a failure to reach agreement that will lead to the exploitation of the mineral wealth of the deep-sea bed by the developed countries, bearing the risks that might attend such unilateral actions.

In the case of energy, a more pragmatic, less ambitious approach may be preferable, under which the United States would take the lead in working out less global schemes, inviting cooperation from those countries ready to enter into agreements by offering to share not only technology, but also the techniques of management and applied science that smaller less developed countries lack. The approach might be to encourage efforts of appropriate technology, emphasizing local needs and resources, supplemented by a willingness to provide capital assistance as well as resources. These arrangements could be made more bilaterally or through such multilateral agencies as the World Bank and the International Monetary Fund.

In this connection, it should also be borne in mind that there is a close connection between energy policy and food policy. While the United States could in time meet its energy needs by relying on a combination of domestic sources of fuel, an effort to rely heavily on biomass sources, especially corn, would almost certainly have negative effects on food production. Since other countries depend heavily on our food exports, we would be putting them in jeopardy. Again, this consideration could serve as a lever in promoting international cooperation. It is not simply a matter of using food supplies to promote good conduct or (as in the case of the embargo against the Soviet Union) to punish bad, but of using food to promote cooperative arrangements. These arrangements could be designed to assist other countries to develop their own sources of energy supply and to increase their self-sufficiency in food resources.

The problems of economic development and modernization in the developing countries have certainly proven difficult to manage, by a variety of social systems. It will be no easy task to surmount all the difficulties. For this reason, it would be best for the United States to single out those cases where development seems most likely to be achieved and to design model experimental approaches upon them, in the hope that they might prove adaptable elsewhere. The Marshall Plan, while more ambitious, could be repeated on a simpler basis, by providing technology and capital to developed economies, coupled with access to our markets. Assisting the energy-poor developing countries will take a more varied approach and will require more active intervention than some of them will find congenial. It is, however, a possible approach to the development of a more comprehensive global regime, which will make manifest our commitment to worldwide applications of democratic principles.

10 Present Needs and Future Prospects

ENERGY AND AMERICAN DEMOCRATIC VALUES

In the past, truly epochal changes in technology and social structure have come about only very gradually. The shift from wood to coal took two hundred years and was the result of a complex combination of factors. Steam engines came into regular and widespread use only a hundred years after the first experimental model was demonstrated. The Industrial Revolution, in large part made possible by the combination of coal as the energy resource and steam as the power system, has still not yet run its course. The same is likely to be true for any social changes that can now be expected as a result of the transition to sustainable forms of energy.

Nevertheless, the cumulative impact of such gradual changes has sometimes led to dramatic transformation of the social order. The Industrial Revolution shifted the focus of human activity from the agricultural village to factories and cities. As coal became more important economically, those who possessed rich resources of coal were able to use these resources to acquire power in international affairs by using them to create an industrial economy and a system of military power built on iron and steel and on warships and armaments. As the first nation to exploit its coal resources in order to industrialize, Great Britain was also the first to achieve world preeminence. Other nations have more recently exploited oil resources and nuclear technology for the same ends.

The British example has been the model for many other nations, including the United States. The energy resources of the new world have made it possible for this relatively new and inexperienced nation to achieve a leading role in the world's economy and eventually to become a great world power. In no country does the historical record offer more dramatic evidence of the important role of energy resources. The social impact of the shift from wood to coal in the nineteenth century and the revolution in ways of life brought about by electrification and the internal combustion engine are evidence from the past. The current and prospective impact of nuclear and solar technologies promises to be equally significant.

Will the future be largely an extrapolation of the industrial past? Or will the achievement of affluence, coupled with growing concern over the effects of depleting scarce resources and increasing risks to health, safety, and ecological systems in order to satisfy the demand for increasing amounts of energy, lead people in the advanced industrial countries to change course? In no country can this question be posed more acutely than in the United States, where "the promise of American life," in Herbert Croly's phrase, has been synonymous with the belief in progress, a belief undergirded by the reality of continued success in industrialization.

It is no exaggeration to say that what is at stake in the present debate over energy policy is the momentum of the process of industrialization, which has been proceeding now for close to 250 years. For all this period, the rate of energy consumption and the rate of economic progress have been closely linked, even when more efficient ways have been found of using energy resources, because these new ways have invariably stimulated new levels of demand for energy services. Can this connection be maintained by the development of greater efficiency and by the development of new sources of energy? Or is the great age of industrialization at an end? Must it be superseded by a new age of frugality, in which nonmaterial objectives replace the drive to industrialize?

Coupled with this long-term issue are more immediate preoccupations. Rising energy prices have been the single greatest source of recent high rates of inflation, and they are at least partly responsible for the slump in industrial productivity. They have been the major cause of the difficulties of the automobile industry and of the severe unemployment those difficulties have entailed. Competition for energy resources also imperils the unity of the NATO alliance and could become the trigger of a major war involving the superpowers. In view of its many serious ramifications, the energy problem must be considered a matter of high social priority.

But just because it is so ramified, just because it is not amenable to quick resolution by a single set of decisions, it poses a major test of the viability of democratic institutions at a time when highly complex technological decisions must be made regularly, and inevitably on the basis of incomplete information. Democratic systems must rely on the ability and willingness of great numbers of people to make informed decisions, both collectively and in their private capacities as businessmen, workers, and consumers. The energy problem will require many decisions that must be made and enforced by governments, but the decisions cannot be reached or made effective unless there is widespread public understanding and support for them. Some of the most important decisions must be made by private organizations and by individuals. For all these decisions to produce a reasonably coherent and coordinated social policy, there must be a common appreciation of the nature of the problem and the general lines of response that must be made to it.

This general understanding does not come about automatically. In 1973, and for some time thereafter, many Americans reacted to the oil embargo and its disruptive effects by refusing to believe that there was a long-term energy problem in prospect. They thought the rise in oil prices had been artificially contrived (as it certainly was) and that it would somehow be rescinded when normal market forces began to be felt (as so far has not happened). The problem, however, is not simply that a free market in energy resources does not exist, but that even a free market could not cope with rising world demand for energy. Steps would have to be taken in any case to moderate demand, increase the availability of alternatives to fossil fuels, and generally manage the transition to more abundant and sustainable sources of supply.

There are some signs, especially the dramatic decline in the consumption of oil, that seem to show not only that consumers are sensitive to market signals but that there is a growing appreciation of the long-term reality of the energy problem. But difficult issues remain to be decided. There will be more debate about nuclear power, though the issue may become somewhat muted as the demand for electricity declines and the rising cost of new nuclear facilities inhibits further investments. Controversy will probably continue to surround nuclear power facilities currently under construction and awaiting operating licenses, as well as the issues of spent fuel storage and radioactive waste disposal. There will be controversy over offshore oil exploration and over the attempt to make more use of coal. There will be continued controversy over the role that government and private industry should play in the effort to manage energy resources and develop alternative technologies. For these issues to be handled successfully, considerable and continued public awareness will be essential.

Will there also need to be a readiness to make major adjustments not only in lifestyle and expectations, but in fundamental values? Traditional attitudes will certainly have to be adjusted in some respects, but so far at least there is little warrant for supposing that radical transformations will need to be made in the social and political system to cope with the impact of the energy problem. Over time, the cumulative effects of incremental changes in lifestyle may amount to a significant alteration in the balance of options available to most people. There may well be less emphasis on the private and more on the public and quasi-public in such areas as housing and transportation. It is conceivable that if the advocates of solar energy are borne out by developments in the marketplace, America may experience a rebirth of community and decentralization. This development, if it comes, will not be experienced rapidly or all at once. If the energy problem persists or becomes worse, there may well be more conflict over the distribution of the social product than this country has previously experienced. Even these controversies, however, will not necessarily call into question the basic structure of American society; nor are they likely to lead to demands for a nationalization of the major energy industries. American values and the American social, economic, and political system need not be fundamentally changed—unless, of course, the problem is so badly managed that some or all of these institutions come to seem inadequate and obsolete. In such a case, the search for scapegoats might easily identify the wrong culprits and provoke a mood of vigilantism that would only make matters worse.

If any of this should come to pass, it is certainly not out of the question that the same tragic polarization that has crippled other democracies in times of crisis could arise in the United States. Whenever a major social problem threatens to overwhelm the fragile resources of democracy, the specter of anarchy and immobilism presents itself at one dark extreme and the specter of repressive and dictatorial movements for law and order arises at the other. The collapse of the brief attempt at constitutional government after the Russian Revolution, and of the Weimar and Spanish republics, are evidence of this fragility under great pressure, and their successor regimes are evidence also of what is likely to happen when democracy collapses. America's energy problem is by no means yet of such proportions as to arouse fears for the survival of the republic. If thoughtful action is taken, it need not become disabling. But there is certainly a risk that if it is not well managed, serious social dislocations could result, which would put the social fabric in jeopardy. More positively, if American democracy can succeed in managing this problem, it will set an example for others of the continuing viability of democratic sytems even in an age of technological complexity. More than any other consideration, it is the need to maintain democracy and to extend its basic principles as a foundation for world order that should guide the American effort to manage the energy problem.

ENERGY CONSERVATION

Among specific policy options, a strong program of energy conservation would contribute significantly to almost all of the values that were outlined in Chapters 4 and 5. A conservation program would reduce global inequities in energy consumption, and it could be designed to include provision for the mitigation of inequitable burdens on low-income families. It would reduce the need for expansion of coal and nuclear technologies with their risks to health, safety, and the environment, and it would itself entail relatively few new risks. It would reduce dependence on foreign oil and the risk of war from international conflict over oil supply. It would facilitate the transition to sustainable sources. Extreme conservation measures would involve some conflict with economic growth and individual freedom, but there is reason to believe policies could be devised that would not place either one in serious jeopardy.

Sudden reductions of energy supply, such as that which followed the OPEC embargo in 1973, are harmful to the economy. But greater efficiency in the use of energy can actually enhance economic growth. The CONAES report suggests that demand could level off and the energy/GNP ratio could be reduced by 50 percent without sacrificing growth in GNP, mainly by greater efficiency in capital equipment and consumer durables. Conservation is not a threat to employment, provided it is effected gradually; it could be a major source of new jobs, widely scattered across the country—more jobs than an equivalent increase in supply. Many forms of conservation result in net capital savings, which could be applied to modernization and productivity improvements. Forty percent of the U.S. capital investment is now allocated to energy production, and according to some projections energy investment could take almost half of all business capital by the end of the century.[1] Measures to slow demand growth would thus help other sectors of the economy that need new capital. The Harvard Business School study concludes that demand could be reduced 30–40 percent—more than all the oil we import—with no decline in economic growth.[2] Three studies published in 1981 hold that a program to reduce demand would consume far less capital than any program designed to boost supply and that government support is warranted to correct market distortions built up by previous supply subsidies.[3]

Oil conservation should have the highest priority because our huge oil imports make us so vulnerable to price rises, shortages, or embargoes. Gasoline for cars and trucks account for 25 percent of all U.S. energy use. Patterns of American life and work were built around the auto in an era of cheap oil. Through regulated prices and highway construction we subsidized cars and trucks, to the detriment of railways, which are generally more energy-efficient. But autos provide job mobility, social status, personal freedom, and convenience, which Americans value highly. Rising gasoline prices have only a modest influence on the number of miles people drive, but they strongly influence the purchase of autos with better fuel economy. The turnover of autos is relatively rapid (half are replaced within five years). Fuel economy standards now require a doubling of fleet-average mileage between 1973 and 1985 (to 27.5 miles per gallon). But the U.S. auto industry, which was slow to adapt, has faced heavy costs in retooling to produce smaller, more efficient autos. Further improvement is technologically feasible; the Volkswagen Diesel Rabbit gets 41 mpg, and the Diesel Turbo 60 mpg.

Proper engine maintenance, stricter enforcement of the 55 mph speed limit, and ride sharing would further reduce fuel use. Riding in a van pool takes one-eighth the energy of a single-occupancy car. Urban buses are rapid and convenient if assigned to express routes and bus lanes; frequent service with smaller buses attracts riders more than lower fares. The electric trolley was once common in U.S. cities, and in some cases trolley routes could be restored. In other industrial societies, and in some U.S. cities, surface transit systems have worked reasonably well, reducing congestion and pollution as well as fuel use. But subways are expensive and save little energy unless they are heavily used.[4]

Industry is the second area in which major energy savings can be effected. Some companies with a strong commitment to energy efficiency have already drastically reduced their demand. Apart from plugging leaks and improving housekeeping to minimize waste, the greatest potential lies in more efficient equipment and industrial processes (especially in steel, aluminum, paper, and chemicals, which are energy-intensive). There are new techniques for waste heat recovery, heat cascading, and the recycling of materials (which often uses far less energy than primary extraction from raw materials). Steam for industrial processes takes 45 percent of all industrial fuel. The cogeneration of steam and electricity is twice as efficient as generating them separately. Twenty-seven percent of West Germany's electricity is produced by industrial firms, half of it by cogeneration. Conversely, heat from central electric stations can be used for district heating of offices and apartment buildings.[5]

A third area for conservation efforts is residential and commercial heating and cooling. In office buildings, better insulation, double-glazing, lower lighting levels, and the recovery of waste heat from ventilated air can cut energy use in half. For new homes, insulation standards, building codes, and loan requirements could lead to far lower heat losses. But housing stock turns over slowly; more rapid savings can come from the "retrofit" of existing houses with weather-stripping and insulation, saving up to 50 percent in heating bills.[6] Since 1978, a 15 percent tax credit (up to $300) has been allowed for residential conservation investment. The Harvard Business School study recommends a 40–50 percent federal subsidy for conservation measures to represent their contribution to national goals, such as reducing our dependence on foreign oil. Some utility companies are taking an active role in promoting conservation, subcontracting and guaranteeing the installation of insulation and arranging loan repayment from the saving in future fuel bills.

A fourth area, electricity demand, is important because electricity is the most capital-intensive form of energy. Regulated rates are currently far below the cost of electricity from new plants; rate increases would have some effect on demand, though the impact would be regressive, as indicated earlier. Lower off-peak rates would produce some shift in demand away from peak-load periods and reduce the generating capacity needed. Water pumped into elevated reservoirs at off-hours can be used to generate electricity during peak periods. Under a 1978 law, utilities are required to buy excess cogenerated electricity from industry. Again, efficiency standards for appliances would cut down on total demand. For example, some air conditioners and refrigerators on the market are twice as efficient as others. New forms of heat pumps can be used efficiently for heating in winter and cooling in summer.

These conservation measures rely on technical changes, economic incentives, and legislated standards, but they would be more effective if accompanied by relatively minor changes in individual behavior. Turning thermostats down in winter and up in summer, turning off unused lights, or using public transportation may seem inconsequential, but many small actions add up to significant savings. Educational programs can bring greater awareness of the importance of conservation and of simple ways to save energy. A study in Three Rivers, New Jersey, showed that home energy consumption fell when special meters were installed to give families rapid feedback of information on their consumption rates. Psychologists have shown the importance of social reinforcement and group support in behavior change.[7] Schipper and Darmstadter write: "We believe that energy will be conserved primarily by technical means, though gradual changes in behavior and lifestyle will help as consumers find themselves as well or better off using less energy."[8]

Neither effective legislation nor behavioral change will be easily achieved, however, because of widespread public complacency about our energy predicament. Moreover, conservation has a diffuse constituency, with few big industrial beneficiaries (except insulation manufacturers). There are powerful institutional forces committed to increases in supply—including groups in industry, labor, and government supporting the use of oil, coal, or nuclear energy—but few organized groups dedicated to the reduction of demand. But there is evidence of growing public support for conservation, reflected in congressional action on auto fuel economy and insulation tax credits.

The measures described so far could be instituted without major changes in lifestyles or value priorities. They depend on technical improvements and economic incentives and require minimal disturbance of existing institutions or behavior. Overall, efficiency in energy use improved by 15 percent from 1973 to 1980, largely as a result of rising prices. Efficiency is a pragmatic and politically acceptable goal on which a wide social consensus is possible. Many observers hold that such measures, along with increases on the supply side, will be enough to balance our energy budget. Anything further might slow economic growth unacceptably, they say, which would be particularly hard on the poor. Any appeal to frugality or sacrifice might lead to public reaction against all conservation proposals. Americans value mobility, material goods, rising standards of living, and convenient products that save time and effort. Such basic values and behavior patterns are deeply engrained and change only very slowly.[9]

But other voices have called for more far-reaching lifestyle and value changes. These people hold that realistic policies must acknowledge environmental and resource constraints and global interdependence, even if the public is reluctant to face them. The social and environmental costs of industrial growth have been increasing, despite improvements in safety and abatement technologies. The growing gap between rich and poor countries is a threat to international stability and peace. Our voracious appetite for oil and other resources is also seen as a violation of global justice. These critics challenge many of the dominant assumptions of American life in the name of both realism and ethics.

A new conservation ethic would be a major departure from prevailing attitudes, but it could draw from several themes in our national heritage. In the nineteenth century, America seemed to be a land of great abundance and seemingly limitless resources. More recently, advertising on TV and other media has further inflated our wants and expectations by identifying happiness and success with ever-increasing consumption. But there are strands in earlier American history that could be recovered. Frugality was once a virtue, wastefulness was condemned, and ingenuity in repairing and reusing things was respected.[10] Again, there are religious themes that provide motives for conservation. In the biblical view, fulfillment is identified with personal existence in community, not with consumption or material possessions. The biblical concern for the basic needs of the neighbor and for justice toward the dispossessed would call into question our disproportionate use of energy.[11] The ecological outlook is another source of concern for conservation. Ecologists have emphasized resource limits, global interdependence, and the indirect costs of energy use. Ecological awareness does not imply a rejection of technology or a romantic return to nature, but rather a recognition of the far-reaching repercussions of our actions. All of these perspectives could contribute to a new conservation ethic.

Moral exhortation and appeals to austerity or sacrifice will probably have little influence today. But a shift in the understanding of human fulfillment can lead to new perceptions and new personal goals. Some people have advocated or adopted simpler lifestyles in which well-being is sought primarily in human relationships, community life, meaningful work, and greater local self-sufficiency. Simplicity is defended not in a spirit of ascetic self-denial, but as part of an alternative vision of the good life. There is a shift here from an industrial to a postindustrial paradigm, not a withdrawal from the world or a return to a pastoral past. While some advocates of alternative lifestyles seem to reject all but the most primitive technologies, most recognize that appropriate kinds of science-based technology are essential if the planet is to support even its present population at an acceptable standard of living.[12]

The Third World, of course, hardly needs to aim for simpler lifestyles. Frugality is not likely to appeal to those still seeking to fill their basic material needs or to escape from poverty and deprivation. Nevertheless, among the relatively well-off in the United States, the voluntary adoption of more frugal living patterns on the basis of new value orientations could provide examples of human fulfillment without ever-increasing energy and resource demands. People might discover that there are satisfying patterns of life compatible with lower levels of energy use. Such examples would be significant if more drastic scarcities are forced on us in the future. In any case, people who voluntarily adopt more frugal lifestyles will contribute to conservation. Such individual action is not incompatible with efforts to adopt effective legislation, economic incentives, and efficient technologies, which are the most promising national paths to energy conservation.

KEEPING OPTIONS OPEN

Conservation is consistent with most of the values discussed in earlier chapters. There is no similar convergence of values in support of any one supply option. We have seen that each of the main energy sources entails uncertain risks and tradeoffs among conflicting values. For example, nuclear energy is clearly preferable to coal with respect to health and environmental impacts in normal operation, but it entails greater risk of catastrophic accidents and is more vulnerable to institutional failures. The prevalence of uncertainties and tradeoffs suggests that it would be desirable to have a diverse mix of sources, which would keep our options open. Such diversity would be consistent with technical and economic prospects, political realism, and the pluralism of values defended in this volume. In our current situation, none of the major supply options should be foreclosed.

Diversity in ecological systems contributes to stability and resiliency. Diverse ecosystems can adapt to new conditions more readily than monocultures. Diverse technologies, too, would offer flexibility in adapting to unexpected developments, as well as allowing alternative ways of meeting different end uses and local conditions. The impacts of the major energy options on health and environment are quite different, and diversity would spread the risks among population groups and regions. Keeping the options open would hedge our bets and provide some insurance against unexpected events. Further knowledge of risks, new safety or pollution control techniques, changing economic costs, or the development of new energy technologies could alter the balance among alternatives.

We have suggested that time is needed to develop advanced converters and alternative fuel cycles that would be more efficient than light-water reactors but less proliferation-prone than breeders. Within a decade or two we should know whether fusion will be technologically and economically feasible. Time is also needed for solar research and development, including cheaper photovoltaic cells, storage systems, and more efficient forms of photosynthesis, which would require less land for biomass conversion. In the meantime, we may gain a better understanding of the SO_2 and CO_2 risks from coal and perhaps achieve improvements in coal technology such as fluidized bed combustion or magnetohydrodynamics. We cannot expect a technical fix to provide cheap risk-free energy, but we can expect that research will yield new knowledge of risks and new improvements in technology, which will affect our policy judgments. Liquid fuels for transportation will be particularly important. If indeed our dependence on foreign oil is dangerous, and increases in domestic oil production are likely to be marginal and temporary (even with enhanced recovery techniques), priority must be given to alternative liquid fuels such as synthetic oil, gasoline or methanol from coal, or methanol or methane gas from biomass. Because of environmental and land-use limitations, biomass can be envisaged as a liquid fuel source only if auto fuel efficiency is greatly increased.

A Swedish study has compared nuclear development (breeders and reprocessing) with solar development (solar heating, photovoltaics, wind, and especially biomass from energy plantations) as long-term substitutes for fossil fuels. The cost of electricity appears lower in the nuclear case, while the cost of heating and transportation fuels appear lower with solar sources; the overall costs are roughly equivalent, but with large uncertainties in both cases. Solar electricity production would be closer to the consumer and would offer more opportunity for cogeneration and district heating. The nuclear future would require strong central authority; the solar future calls for more planning by local authorities but also requires extensive national planning to integrate land use, energy distribution and energy end use. The study outlines a strategy that does not foreclose either the nuclear or the solar option. It recommends policies to strengthen local governments and businesses in energy fields (including conservation, distribution, and district heating). Another feature is the design of flexible end-use patterns that could be adapted to alternative supply sources in the future. Effective local and national planning would be needed to ensure that land is available for reactor sites, biomass plantations, and solar heating stations, if intermediate-term flexibility were accepted as a national goal. But the study warns that the nuclear industry has greater economic power and institutional momentum than solar industries and will tend to dominate choices unless solar options are deliberately cultivated.[13]

A mix of energy sources appears most likely to be acceptable to an American public that is divided about nuclear energy. In one 1980 survey, 20 percent of respondents said that all existing nuclear plants should be shut down, 47 percent said that we should use those already built but build no additional ones, and 23 percent said that building nuclear plants should continue. When people were asked to choose from a list of energy options those two or three on which we should concentrate in looking ahead to the year 2000, solar energy was chosen most often (61 percent) and nuclear energy least (23 percent), with coal, conservation, water power, oil and natural gas, and synfuels (in that order) in between.[14] Another 1980 survey found that 4 percent of respondents said they were active in the antinuclear movement, 29 percent sympathetic with it, and 26 percent unsympathetic, but 41 percent were neutral or undecided.[15]

Kasperson, et al. suggest that both sides in the nuclear debate espouse values that appeal to the uncommitted public. Nuclear advocates stress economic growth, environmental protection, the risks of energy shortages, and an optimistic view of technology. Nuclear critics stress participation in decisions, the risks of catastrophic accidents, and the institutional difficulties in controlling large-scale technologies. Kasperson, et al. hold that neither set of values is predominant in American society and a clear victory for either side is unlikely. Better information may aid the resolution of some aspects of the debate but will not resolve the underlying value conflicts. Future events may influence the balance but will probably not be decisive; a cutoff of Middle Eastern oil, for example, would favor the nuclear advocates, while a major reactor accident would favor the critics. These authors conclude that the broad majority is likely to favor a compromise that avoids rapid expansion of nuclear plants but keeps the nuclear option open while other alternatives are explored.[16]

A study sponsored by Resources for the Future concludes that the values of economic growth and preservation of environment and health are not incompatible but will require careful planning and some compromise. To secure the cooperation of environmentalists, and to protect environmental values, regulations for safety and pollution control will have to be stronger than the nuclear industry or the coal industry would like. But environmentalists will also have to give greater recognition to the importance of adequate energy supplies for a healthy economy; they will have to acknowledge that it is very unlikely that conservation and solar energy alone could replace foreign oil in this century without severe economic disruption.[17]

Because the tradeoffs among conflicting values involve value judgments as well as scientific judgments, they should be made through political processes and not by technical experts alone. Energy policy decisions entail differing distributions of costs, risks, and benefits among various population segments. Interest-group politics will probably continue to be dominant in the United States, and it is unlikely that a strong social consensus will emerge. In a pluralistic society with diverse regional interests, policies are more likely to be formed from tradeoffs and coalitions than from a clear consensus. We can hope, however, that there will be a greater awareness of common interests, a greater willingness to accept compromises for the public good, and a stronger dedication to democratic processes of decision making. Only then can we find our way through the complex issues of energy policy without abandoning the basic values of our national heritage.

Notes

CHAPTER 1: ENERGY AND THE RISE OF AMERICAN INDUSTRIAL SOCIETY

1. See M.D. Wyatt, *Industrial Arts of the Nineteenth Century at the Great Exhibition, 1851* (London, 1851–1853); Cristopher Hobhouse, *1851 and the Crystal Palace* (New York: E.P. Dutton, 1950); C.R. Fay, *Palace of Industry, 1951* (Cambridge: The University Press, 1951).
2. See John U. Nef, *The Conquest of the Material World* (Chicago: The University of Chicago Press, 1964).
3. See David Landes, *Unbound Prometheus* (London: Cambridge University Press, 1969).
4. See C.E. Ayres, *Toward a Reasonable Society* (Austin, Tex: University of Texas Press, 1961), pp. 229–46.
5. See the perspective suggested in William H. McNeill, *The Rise of the West* (Chicago: The University of Chicago Press, 1963).
6. Melvin Kranzberg and Joseph Gies, *By the Sweat of Thy Brow* (New York: G.P. Putnam's Sons, 1975), pp. 28–38, 45–47.
7. Lynn White, Jr., *Medieval Technology and Social Change* (New York: Oxford University Press, 1962), pp. 79ff.
8. Ibid., pp. 57–69.
9. Kranzberg and Gies, *Sweat of Thy Brow,* pp. 79ff.
10. See William Woodruff, *Impact of Western Man: A Study of Europe's Role in the World Economy, 1750–1960* (New York: St. Martin's Press, 1966).
11. Abbott Payson Usher, "The Textile Industry, 1750–1830," in *Technology in Western Civilization,* ed. Melvin W. Kranzberg and Carroll W. Pursell, Jr. (New York: Oxford University Press, 1967), vol. I, chap. 14.
12. John Brinkerhoff Jackson, *American Space* (New York: W.W. Norton, 1972), pp. 231–40.
13. Henry Adams, *History of the United States During the Administrations of Jefferson and Adams* (New York: Charles Scribner's Sons, 1890), vol. I, pp. 16–17.
14. See Curtis P. Nettles, *The Emergence of a National Economy* (New York: Henry Holt, 1962).

15. Fernand Braudel, *Capitalism and Material Life, 1400–1800* (New York: Harper & Row, 1974), pp. ix–xiv, 325–74.

16. See the essays edited by Brook Hindle in *America's Wooden Age* (Tarrytown, N.Y.: Sleepy Hollow Press, 1979), and *Material Culture of the Wooden Age* (Tarrytown, N.Y.: Sleepy Hollow Press, 1980).

17. See the useful discussion in Adrienne Koch, ed., *The American Enlightenment* (New York: George Braziller, 1965), pp. 45, 281, 347.

18. Russel Blaine Nye, *The Cultural Life of the New Nation, 1776–1830* (New York: Harper & Row, 1960), p. 44.

19. See Koch, *American Enlightenment,* pp. 46ff, 106, 119–21, 187–88.

20. See Henry Steele Commager, *The Empire of Reason* (New York: Anchor Press/Doubleday, 1978), pp. 33, 40, 44ff, 79, 195, 257.

21. Marvin Fisher, *Workshops in the Wilderness* (New York: Oxford University Press, 1967).

22. J. Hector St. John de Crevecoeur, *Letters from an American Farmer* (New York: E.P. Dutton, 1957; first published 1782), p. 61.

23. Quoted in Fisher, *Workshops* p. 87.

24. Quoted in Koch, *American Enlightenment* p. 393.

25. See especially John F. Kasson, *Civilizing the Machine* (New York: Penguin Books, 1976, 1977), pp. 14–52; Leo Marx, *The Machine in the Garden: Technology and the Pastoral Ideal in America* (New York: Oxford University Press, 1964), pp. 150–69.

26. Quoted in Koch, *American Enlightenment* p. 633.

27. Quoted in Adams, *History,* vol. I, p. 43. See the interpretation by Richard A. Bartlett, *The New Country: A Social History of the American Frontier, 1776–1890* (New York: Oxford University Press, 1974), pp. 175–211.

28. Quoted in Adams, *History,* vol, I, p. 42. See Reginald Horsman, *The Frontier in the Formative Years, 1783–1815* (New York: Holt, Rinehart and Winston, 1970), pp. 111–19.

29. R.G. Lillard, *The Great Forest* (New York: Alfred A. Knopf, 1948).

30. Quoted in H.J. Habakkuk, *American and British Technoloyy in the Nineteenth Century* (London: Cambridge University Press, 1962), p. 12.

31. Quoted in Lewis C. Gray, *History of Agriculture in the Southern United States to 1860* (Gloucester, Mass.: Peter Smith, 1958), vol. I, p. 452.

32. See John T. Schlebecker, *Whereby We Thrive: A History of American Farming, 1607–1972* (Ames, Iowa: The Iowa State University Press, 1975).

33. See Clarence Danhof, *Change in Agriculture, 1820–1870* (Cambridge, Mass.: Harvard University Press, 1969).

34. See Nathan Rosenberg, *Technology and American Economic Growth* (White Plains, N.Y.: M.E. Sharpe, 1972), pp. 89n, 142n.

35. Joseph M. Petulla, *American Environmental History* (San Francisco: Boyd and Fraser, 1977), pp. 105–6.

36. Adams, *History,* vol. I, p. 70.

37. Ibid., pp. 181–82.

38. See Greenville and Dorothy Blake, *Oliver Evans: A Chronicle of Early American Engineering* (Philadelphia: Historical Society of Pennsylvania, 1935).

39. Petulla, *Environmental History,* pp. 118–19.

40. See John W. Oliver, *History of American Technology* (New York: The Ronald Press, 1956).

41. See Constance McL. Green, *Eli Whitney and the Birth of American Technology* (Boston: Little, Brown and Company, 1956).

42. See Stuart Bruchey, *The Roots of American Economic Growth, 1607–1861* (New York: Harper & Row, 1965, 1968), pp. 89–90.

43. See Norman J. Ware, *The Industrial Worker, 1840–1860* (Boston: Houghton Mifflin, 1924).

44. David J. Jeremy, "Innovation in American Textile Technology During the Early 19th Century," *Technology and Culture* 14 (1973): 40–76.

45. Kranzberg and Gies, *Sweat of Thy Brow* pp. 92–99.

46. See Benjamin P. Johnson, "America at the Crystal Palace, 1851," in *Readings in Technology and American Life* ed. Carroll W. Pursell, Jr. (New York: Oxford University Press, 1969), pp. 96–101.

47. L.C. Hunter, *Steamboats on the Western Rivers* (Cambridge, Mass.: Harvard University Press, 1949).

48. George Rogers Taylor, *The Transportation Revolution, 1815–1860* (New York: Holt, Rinehart and Winston, 1951).

49. See the very important interpretation in Charles Moraze, *The Triumph of the Middle Class* (London: Weidenfeld and Micolson, 1966).

50. See, for example, Oscar Osburn Winther, *The Transportation Frontier: Trans-Mississippi West, 1865–1890* (New York: Holt, Rinehart and Winston, 1964), pp. 92ff.

51. See Alfred D. Chandler, *The Railroads: The Nation's First Big Business* (New York: Harcourt Brace and World, 1965).

52. See Leo Marx, *Machine in the Garden,* pp. 194–209.

53. Robert L. Heilbroner, *The Economic Transformation of America* (New York: Harcourt Brace Jovanovich, 1977), p. 54.

54. See Petulla, *Environmental History,* pp. 151–52.

55. Peter Temin, *Iron and Steel in Nineteenth Century America* (Cambridge, Mass.: Harvard University Press, 1964).

56. See Petulla, *Environmental History,* pp. 178–82.

57. *Autobiography of Andrew Carnegie* (Boston: Houghton Mifflin, 1920), p. 227.

58. Heilbroner, *Economic Transformation,* p. 90.

59. Ibid., pp. 95–96.

60. Sam H. Schurr and Bruce C. Netschert, *Energy in the American Economy, 1850–1975* (Baltimore: The Johns Hopkins Press, 1960), pp. 325–45.

61. Heilbroner, *Economic Transformation,* pp. 67ff.

62. Conveniently available in Thomas Parke Hughes, ed., *Changing Attitudes Toward American Technology* (New York: Harper & Row, 1975), pp. 166–75.

63. The title of Howard Mumford Jones' immensely insightful book, *The Age of Energy: Varieties of American Experience, 1865–1915* (New York: The Viking Press, 1970–71), esp. pp. 100–78.

64. Henry Steele Commager, *The American Mind* (New Haven, Conn.: Yale

University Press, 1954), p. 13.

65. Charles Beard and Mary Beard, *The Rise of American Civilization* (New York: Macmillan, 1933) vol. II, p. 395.

66. The most useful review of Carnegie's philosophy and the resulting controversy is still Gail Kennedy, *The Gospel of Wealth* (Boston: D.C. Heath, 1949).

67. David T. Bazelon, "The New Factor in American Society," in *Environment and Change: The Next Fifty Years,* ed. William R. Ewald, Jr. (Bloomington, Ind.: Indiana University Press, 1968), p. 267.

68. Kranzberg and Gies, *Sweat of Thy Brow,* pp. 151ff.

69. Heilbroner, *Economic Transformation,* p. 102. See also S.A. Lakoff, ed., *Private Government: Introductory Readings* (Glenview, Ill.: Scott, Foresman, 1973).

70. Bazelon, "New Factor," p. 267.

71. U.S. Senate, Committee on Education and Labor, *Report on the Relations Between Labor and Capital* (Washington, D.C.: U.S. Government Printing Office, 1885), vol. I, p. 473.

72. One of the most perceptive accounts is in Frederic Cople Jaher, *Doubters and Dissenters* (London: Free Press, 1964).

73. Heilbroner, *Economic Transformation,* pp. 143–45, ties material abundance directly to industrial productivity.

74. Richard M. Abrams, *The Burdens of Progress, 1900–1929* (Glenview, Ill.: Scott, Foresman, 1978), p. iv.

75. Quoted in ibid., p. 2.

76. Ibid., p. 8.

CHAPTER 2: ENERGY AND ABUNDANCE: ADVANCE AND RETREAT

1. The best statistical and historical analysis is still in Schurr and Netschert, *Energy,* where the indexed listing on "electricity" and "electrification" covers more than two columns.

2. See H.E. Passer, *The Electrical Manufacturers, 1857–1900* (Cambridge, Mass.: Harvard University Press, 1953).

3. *Edison: Lighting A Revolution. The Beginning of Electric Power* (Washington, DC: National Museum of History and Technology, 1979).

4. Thomas Parke Hughes, "The Electrification of America: The System Builders," *Technology and Culture* 20 (January 1979): 124–61.

5. "Harnessing a Monument: The Power of Niagara," *EPRI Journal IV* (March 1979): 31–39.

6. See Passer, *Electrical Manufacturers.*

7. *Chester T. Crowell, "Nine Slaves for Each Citizen," Saturday Evening Post* (November 6, 1926).

8. The best statistical and historical analysis is in Schurr and Netschert, *Energy,* where the indexed listing on "coal" covers more than four columns. However, the study ends in 1955.

9. See note 1.

10. *Edison: Lighting a Revolution,* pp. 81–85.

11. Quoted in Abrams, *Burdens of Progress,* pp. 166–67.

12. The most useful study is still P.H. Giddens, *The Birth of the Oil Industry* (New York: Macmillan, 1938).

13. Petulla, *Environmental History,* pp. 295–98, 343–44.

14. Schurr and Netschert, *Energy,* pp. 96–124.

15. See Daniel Ford, "Three Mile Island" *The New Yorker,* April 6, 1981; April 13, 1981.

16. Richard Rhodes, "A Demonstration at Shippingport: Coming on Line," *American Heritage* 32 (June/July 1981): 66.

17. Quoted in ibid., p. 68.

18. Ibid., p. 69.

19. Quoted in ibid., p. 70.

20. Ibid., p. 72.

21. Quoted in ibid.

22. See *Report of The President's Commission on The Accident at Three Mile Island* (New York: Pergamon Press, 1979).

23. David J. Rose, "Energy and History," *American Heritage* 32 (June/July 1981): 79–80.

24. David M. Potter, *People of Plenty: Economic Abundance and the American Character* (Chicago: The University of Chicago Press, 1950), esp. pp. 111–41.

25. Scherr and Netschert, *Energy,* p. 77.

26. See the useful summary in W. Elliot Brownlee, *Dynamics of Ascent: A History of the American Economy* (New York: Alfred A. Knopf, 1974), esp. chap. 8, pp. 189–210.

27. Scherr and Netschert, *Energy,* p. 47.

28. Heilbroner, *Economic Transformation,* p. 211.

29. The literature here is large: see, for example, Jacques Ellul, *The Technological Society* (New York: Random House/Vintage, 1964); Lewis Mumford, *The Myth of the Machine* (New York: Harcourt Brace Jovanovich, 1967, 1970); C.E. Black, *Dynamics of Modernization* (New York: Harper & Row, 1966).

30. Bertrand de Jouvenel, "On Attending to the Future," in *Environment and Change,* ed. Ewald p. 29.

31. See Roderick Nash, *Wilderness and the American Mind* (New Haven: Yale University Press, 1967).

32. See, for example, Hans Huth, *Nature and the American* (Berkeley, Calif.: The University of California Press, 1957).

33. Among recent books, see especially George Steiner, *In Bluebeard's Castle* (New Haven, Conn.: Yale University Press, 1971).

34. See the updated discussion in Ian G. Barbour, *Technology, Environment, and Human Values* (New York: Praeger, 1980), pp. 41–49.

35. Nash, *Wilderness, pp. 132–33.*

36. *The classic analysis remains Samuel P. Hays, Conservation and the Gospel of Efficiency* (Cambridge, Mass.: Harvard University Press, 1959).

37. See John Ise, *The United States Oil Policy* (New Haven, Conn.: Yale University Press, 1926).

38. See Petulla, *Environmental History,* pp. 280–81.

39. See Nash, *Wilderness,* for a discussion of Hetch-Hetchy that makes the event a crucial turning point in the history of the preservation-conservation conflict.

40. Frank Graham, Jr., *Since Silent Spring* (Boston: Houghton Mifflin, 1970).

41. Richard H.K. Vietor, *Environmental Politics and the Coal Coalition* (College Station, Tex.: Texas A & M University Press, 1980).

42. See such risk analysis studies as Baruch Fischoff, et al., *Approaches to Acceptable Risk: A Critical Guide* (Oak Ridge, Tenn.: U.S. Nuclear Regulatory Commission, Oak Ridge National Laboratory, 1980), pp. 27, 79ff.

CHAPTER 3: THE DEBATE OVER ENERGY POLICY

1. Charles Proteus Steinmetz's paper was published in *Transactions of the American Institute of Electrical Engineers,* vol. XXXVI, June 27, 1918, pp. 985–91.

2. Robert Gillette, "Oil and Gas Resources: Did USGS Gush Too High?" in *Energy: Use, Conservation and Supply,* ed. P. Abelson (Washington, D.C.: American Association for the Advancement of Science, 1974), p. 69.

3. For a fuller review of the findings of the Paley Commission see Craufurd D. Goodwin, ed., *Energy Policy in Perspective* (Washington, D.C.: Brookings, 1981), pp. 53–60.

4. Ibid., p. 398.

5. The comparison has been drawn by Andrew Hacker in "Nuclear Power and the Hamilton-Jefferson Debate," *Electric Perspectives* (Summer 1980): 1–13, and John Agresto, "Hamilton vs. Jefferson: This Time It's Energy," *The New York Times,* August 23, 1981, p. E21.

6. Amory Lovins, "Energy Strategy: The Road Not Taken?", *Foreign Affairs* (October 1976): 65–96.

7. Hugh Nash, ed., *The Energy Controversy: Soft Path Questions and Answers By Amory Lovins and His Critics* (San Francisco: Friends of the Earth, 1979), p. 249. This book assembles commentaries critical of Lovins's views and Lovins's responses to those criticisms.

8. Amory B. Lovins, *Soft Energy Paths: Toward a Durable Peace* (San Francisco: Friends of the Earth International, 1977), p. 11.

9. Ibid., p. 14.

10. Ibid., p. 55.

11. Nash, ed., *Energy Controversy,* pp. 15–16.

12. Ibid., p. 19.

13. Lovins, *Soft Energy Paths,* p. 10.

14. Ibid., p. 14.

15. Ibid., p. 40.

16. Ibid.

17. Ibid., p. 42n.

18. Ibid., p. 53, pp. 171–97, and Nash, ed., *Energy Controversy,* pp. 29, 205.

19. Lovins, *Soft Energy Paths,* p. 19.
20. Ibid., p. 57.
21. Ibid., p. 151, and Nash, ed., *Energy Controversy,* p. 212.
22. Nash, ed., *Energy Controversy,* p. 168.
23. Ibid., p. 215.
24. The criticism is put by Harry Perry and Sally Streiter in Nash, ed., *Energy Controversy,* pp. 334–35.
25. Lovins, *Soft Energy Paths,* p. 12.
26. Nash, ed., *Energy Controversy,* p. 246.
27. Sheldon Butt, in ibid., p. 214.
28. Lovins, *Soft Energy Paths,* p. 168.
29. Nash, ed., *Energy Controversy,* p. 206.
30. Ibid., pp. 206–7.
31. Ibid., p. 207.
32. Thomas Jefferson, *Notes on the State of Virginia, 1785* (New York: Harper Torchbooks, 1964), p. 140.
33. See especially Louis Hartz, *The Liberal Tradition in America* (New York: Harcourt Brace, 1955).
34. See for this point of view: Barry Commoner, *The Poverty of Power* (New York: Alfred A. Knopf, 1976); S. David Freeman, *Energy, the New Era* (New York: Vintage Books, 1974).
35. Robert Stobaugh and Daniel Yergin, eds., *Energy Future* (New York: Ballantine Books, 1979), p. 278.

CHAPTER 4: VALUES IN CONFLICT

1. Kurt Baier and Nicholas Rescher, eds., *Values and the Future* (New York: The Free Press, 1969), chaps. 1 and 2; see also Barbour, *Technology,* chap. 4.
2. Milton Rokeach, *The Nature of Human Values* (New York: The Free Press, 1973); Daniel Yankelovich, *New Rules* (New York: Random House, 1981).
3. P.H. Partridge, "Freedom" in *Encyclopedia of Philosophy,* ed. Paul Edwards (New York: Macmillan, 1967); Joel Feinberg, *Social Philosophy* (Englewood Cliffs, N.J.: Prentice-Hall, 1973), chap. 1; Gerald MacCallum, "Negative and Positive Freedom," *Philosophical Rev.* 76 (1967):312–34.
4. David Rossin, "The Soft Energy Path: Where Does it Really Lead," *The Futurist* 14 (1980):57–63; Rossin, "Centralization, Decentralization and Polarization" in *Energy: The Ethical Issues* (Springfield, Ohio: Ohio Institute for Appropriate Technology, 1978).
5. John Palmer, John Todd, and Howard Tuckman, "The Distributional Impact of Higher Energy Prices," *Public Policy* 24 (Fall 1976):545–68; Denton Morrison, "Equity Impacts of Some Major Energy Alternatives," in *Energy Policy in the U.S.: Social and Behavioral Dimensions,* ed. Seymour Warkhov (New York: Praeger, 1978).
6. Joel Primack and Frank von Hippel, *Advice and Dissent* (New York: Basic Books, 1974); Thomas J. Kuehn and Alan L. Porter, eds., *Science, Technology and*

National Policy (Ithaca, N.Y.: Cornell University Press, 1981); Milton R. Wessel, *Science and Conscience* (New York: Columbia University Press, 1980).

7. Steven Del Sesto, *Science, Politics and Controversy: Civilian Nuclear Power in the U.S., 1946–1974* (Boulder, Colo.: Westview, 1979); Elizabeth Rolph, *Nuclear Power and the Public Safety: A Study in Regulation* (Lexington, Mass.: Lexington, 1979).

8. Roger Kasperson, et al., "Public Opposition to Nuclear Energy: Retrospect and Prospect," *Science, Technology and Human Values* 5 (Spring 1980): 11–23; Julia Bickerstaffe and David Pearce, "Can There Be a Consensus on Nuclear Power?" *Social Studies of Science* 10 (1980):309–44; Stanley Nealey, *Nuclear Power and the Public* (Lexington, Mass.: Lexington, 1980).

9. Michael Crozier, et al., *The Crisis of Democracy* (New York: New York University Press, 1975), chaps. 3 and 5; Daniel Bell, *The Cultural Contradictions of Capitalism* (New York: Basic Books, 1976), chap. 6.

10. Dorothy Nelkin and Susan Fallows, "The Evolution of the Nuclear Debate: The Role of Public Participation," *Annual Review of Energy* 3 (1978):275–312; Nancy Abrams and Joel Primack, "The Public and Technological Decisions," *Bulletin of the Atomic Scientists* 36 (June, 1980):44–48.

11. Dorothy Nelkin and Michael Pollack, *The Atom Besieged: Extra-parliamentary Dissent in France and Germany* (Cambridge, Mass.: MIT Press, 1980).

12. Russell Ayres, "The Policing of Plutonium: The Civil Liberties Fallout," *Harvard Civil Rights-Civil Liberties Law Rev.* 10 (1975):369–443; Paul Sieghart, "Guarding Nuclear Materials and Civil Liberties," *Bulletin of the Atomic Scientists* 36 (May, 1980):32–34.

13. Mason Willrich and T.B. Taylor, *Nuclear Theft: Risks and Safeguards* (Cambridge, Mass.: Ballinger, 1974).

14. Lovins, *Soft Energy Paths;* Denis Hayes, *Rays of Hope: The Transition to a Post-Petroleum World* (New York: W.W. Norton, 1977); Kirkpatrick Sale, *Human Scale* (New York: Coward, McCann, Geoghegan, 1980).

15. See replies to Lovins by George Pickering and others in *Alternative Long-Range Energy Strategies: Additional Appendices* (Select Committee on Small Business, U.S. Senate, 1977); also Samuel Florman, "Small is Dubious," *Harper's* (August 1977): 10–12.

16. Committee on Nuclear and Alternative Energy Systems (CONAES), National Academy of Sciences, *Energy in Transition: 1985–2010* (San Francisco: W.H. Freeman, 1980), p. 97.

17. On the pros and cons of decentralization, see chapters by Langdon Winner and Harvey Brooks in *Appropriate Technology and Social Values: A Critical Appraisal,* ed. Franklin Long and Alexandra Oleson (Cambridge, Mass.: Ballinger, 1980).

18. Arthur Okun, *Equality and Efficiency* (Washington, D.C.: Brookings Institution, 1975); see also Hugo A. Bedau, ed., *Justice and Equality* (Englewood Cliffs, N.J.: Prentice-Hall, 1971).

19. John Rawls, *A Theory of Justice* (Cambridge, Mass.: Harvard University Press, 1971).

20. Robert Nozick, *Anarchy, State and Utopia* (New York: Basic Books, 1974); Brian Barry, *The Liberal Theory of Justice* (Oxford: Oxford University Press, 1973);

Norman Daniels, ed., *Reading Rawls: Critical Studies in a Theory of Justice* (New York: Basic Books, 1974).

21. George W. Forell, *Christian Social Teachings* (Minneapolis: Augsburg, 1971); John Haughey, ed., *The Faith that Does Justice* (New York: Paulist Press, 1971).

22. Gallup poll reported in *Christian Century,* May 13, 1981, p. 535.

23. For energy statements by the National Council of Churches, see Dieter Hessel, ed., *Energy and Ethics* (New York: Friendship Press, 1979); on the World Council of Churches, see Paul Abrecht, ed., *Faith and Science in an Unjust World,* (Philadelphia: Fortress Press, 1980), vol. 2, chap. 6.

24. Dorothy Nelkin, ed., *Controversy: Politics of Technical Decisions* (Beverley Hills, Calif.: Sage, 1979), part II; Rayna Green, et al., *Report of a Conference on Energy Resource Development and Indian Lands* (Washington, D.C.: American Association for the Advancement of Science, 1978); Dorothy Nelkin, "Native Americans and Nuclear Power," *Science, Technology and Human Values* 6 (Spring 1981):2–13.

25. *U.S. Coal Development: Promises and Uncertainties* (General Accounting Office, 1977).

26. Roger Kasperson, "Confronting Equity in Radioactive Waste Management," paper presented at meeting of American Association for the Advancement of Science (January 1981).

27. William Ramsey, *Unpaid Costs of Electrical Energy* (Baltimore: The Johns Hopkins University Press, 1979); *The Direct Use of Coal* (Office of Technology Assessment, 1979).

28. William Lowrance, *Of Acceptable Risk* (Los Altos, Calif.: William Kaufman, 1976); *Toxic Substances: Decisions and Values,* vols. 1 and 2 (Washington, D.C.: Technical Information Project, 1979).

29. K.S. Shrader-Frechette, *Nuclear Power and Public Policy: The Social and Ethical Problems of Fission Technology* (Boston: Reidel, 1980), chap. 2.

30. Report of Fuel Oil Marketing Advisory Committee, Department of Energy, cited in *Minneapolis Tribune,* July 22, 1980. See also Bob Swierczek and David Tyler, "Energy and the Poor," *Christianity and Crisis* 38 (1978): 242–44; Eugene Grier, *Colder... Darker: The Energy Crisis and Low-Income Americans* (Washington, D.C.: Center for Metropolitan Studies, 1977).

31. William Baumol, "Environmental Protection and Income Distribution," in *Benefit-Cost and Policy Analysis 1974,* ed. Richard Zeckhauser, et al. (Chicago: Aldine, 1975).

32. Robert Stobaugh and Daniel Yergin, eds., *Energy Future,* revised ed. (New York: Ballantine Books, 1980), p. 223.

33. Lenneal Henderson, "Energy Policy and Social Equity" in *New Dimensions to Energy Policy,* ed. Robert Lawrence (Lexington, Mass.: Lexington, 1979).

34. Charles R. Beitz, *Political Theory and International Relations* (Princeton, N.J.: Princeton University Press, 1979); see also Robert Amdur, "Rawls' Theory of Justice: Domestic and International Perspectives," *World Politics* 29 (1977):438–61.

35. 1979 UNESCO figures, cited in *Anticipation* No. 28 (December 1980): 48.

36. Joseph Egan and Shem Arungu-Olende, "Nuclear Power for the Third World?" *Technology Review* 82 (May 1980):46–55; Jorge Sabato and Jairam Ramash,

"Atoms for the Third World," *Bulletin of the Atomic Scientists* 36 (March 1980): 36–43.

37. Norman Brown and James Howe, "Solar Energy for Village Development," *Science* 199 (1978):651–56; A.K.N. Reddy, "Energy Options for the Third World," *Bulletin of the Atomic Scientists* 34 (May, 1978):28–33; Thomas Hoffman and Brian Johnson, *Energy Cooperation: A New Strategy Toward the Third World* (Cambridge, Mass.: Ballinger, 1980); Norman Brown, "Renewable Energy Resources for Developing Countries," *Annual Review of Energy* 5 (1980):389–413.

38. See J.J.C. Smart and Bernard Williams, *Utilitarianism: For and Against* (New York: Cambridge University Press, 1973); William Frankena, *Ethics,* 2nd ed. (Englewood Cliffs, N.J.: Prentice-Hall, 1973), chaps. 2 and 3.

39. Matthew 25:35.

40. L. Harold DeWolf, *Responsible Freedom* (New York: Harper & Row, 1971); Forell, *Christian Social Teachings.*

41. Donella Meadows et al., *The Limits to Growth* (New York: Universe, 1972); see also Dennis Meadows, ed. *Alternatives to Growth—I* (Cambridge, Mass.: Ballinger, 1977).

42. Mihajlo Mesarovic and Eduard Pestel, *Mankind at the Turning Point* (New York: E.P. Dutton, 1974); Robert Stivers, *The Sustainable Society* (Philadelphia: Westminster, 1976).

43. Margaret Maxey, "Nuclear Energy Politics: Moralism versus Ethics," (Washington, D.C.: Ethics and Public Policy Center, 1977); Frederick Carney, "An Ethical Analysis of Nuclear Power," *America,* September 6, 1980, pp. 86–92.

44. *The New York Times,* April 6, 1980.

45. Eliot Marshall, "Energy Forecasts: Sinking to New Lows," *Science* 208 (1980):1353–56; Energy Study Group (Hans Landsberg, chairman), *Energy: The Next 20 Years* (Cambridge, Mass.: Ballinger, 1979).

46. J. Darmstadter, et al., *How Industrial Societies Use Energy: A Comparative Analysis* (Baltimore: The Johns Hopkins University Press, 1978).

47. CONAES, *Energy in Transition;* also Demand and Conservation Panel, CONAES, "U.S. Energy Demand: Some Low Energy Futures," *Science* 200 (1978): 142–52; Harvey Brooks, "Energy: A Summary of the CONAES Report," *Bulletin of the Atomic Scientists* 36 (Feb., 1980):23–30.

48. Dorothy Nelkin, "Labor and Nuclear Power," *Environment* 22 (March 1980): 6–13.

49. *Environmental Quality 1978* (Council on Environmental Quality, 1978), pp. 431–32.

50. *Creating Jobs Through Energy Policy,* hearings before the Joint Economic Committee, U.S. Congress (March 15–16, 1978); *Jobs, Energy and the Environment* (Washington, D.C.: Environmentalists for Full Employment, 1979).

51. Department of Labor figures cited in *Minneapolis Tribune,* March 21, 1980.

52. Harry Otway, et al., "Nuclear Power: The Question of Public Acceptance," *Futures* 10 (1978):109–18.

53. Lee Schipper and Allen Lichtenberger, "Efficient Energy Use and Well-Being: The Swedish Example," *Science* 194 (1976):1001–13; Joy Dunkerley, ed., *International Comparisons of Energy Consumption* (Baltimore: The Johns Hopkins University Press, 1978).

54. Laura Nader and Stephen Beckerman, "Energy as It Relates to Quality and Style of Life," *Annual Revue of Energy* 3 (1978): 1–28; *Energy Choices in a Democratic Society,* CONAES Supporting Paper No. 7 (National Academy of Sciences, 1980).

55. Daniel Yankelovich and Bernard Lefkowitz, "The Public Debate on Growth: Preparing for Resolution," *Technological Forecasting and Social Change* 17 (1980): 95–140.

CHAPTER 5: UNCERTAIN RISKS

1. Nuclear Energy Policy Study Group (the Ford-Mitre Report), *Nuclear Power: Issues and Choices* (Cambridge, Mass.: Ballinger, 1977), chap. 1; see also David Rose, et al., "Nuclear Power—Compared to What?" *American Scientist* 64 (1976):291–300; Christopher Hohenemser, et al., "The Distrust of Nuclear Power," *Science* 196 (1977):25–34.

2. Ramsey, *Unpaid Costs;* Stephen Barrager, et al., *The Economic and Social Costs of Coal and Nuclear Electric Generation* (National Science Foundation, 1976); *Risks Associated with Nuclear Power: A Critical Review of the Literature* (National Academy of Sciences, 1979).

3. Lowrance, *Acceptable Risk; Decision Making for Regulating Chemicals in the Environment* (National Academy of Sciences, 1975); British Council for Science and Society, *The Acceptability of Risk* (Chichester, England: Barry Rose, 1977).

4. *Reactor Safety Study* (WASH-1400) (Nuclear Regulatory Commission, 1975); "Report to the American Physical Society by the Study Group on Light-Water Reactor Safety," *Review of Modern Physics* 47, supp. no. 1 (1975).

5. Allan Mazur, "Disputes between Experts," *Minerva* 11 (1973):243–62; Sanford A. Lakoff, "Scientists, Technologists and Political Power," in *Scientists, Technology and Society,* ed. Ina Spiegel-Rösing and Derek de Solla Price (Beverly Hills, Calif.: Sage, 1977); Danield D. McCracken, *Public Policy and the Expert: Ethical Problems of the Witness* (New York: Council on Religion and International Affairs, 1971); Nelkin, ed., *Controversy;* D. Robbins and R. Johnston, "The Role of Cognitive and Occupational Differentiation in Scientific Controversies," *Social Studies of Science* 6 (1976):349–68.

6. Barry Casper, "Technology Policy and Democracy," *Science* 194 (1976): 29–35.

7. Wilma McCarey, "Pesticide Regulation: Risk Assessment and Burden of Proof," *George Washington Law Revue* 45 (1977):1066–94; Paul Portney, "Toxic Substances and the Protection of Human Health," in *Current Issues in Environmental Policy,* ed, P. Portney (Baltimore: The Johns Hopkins University Press, 1978); Norman Vig, "Environmental Decision-Making in the Lower Courts: The Reserve Mining Case," in *Energy and Environmental Issues: The Making and Implementation of Public Policy,* ed. Michael Steinman (Lexington, Mass.: Lexington, 1979).

8. Nuclear Energy Policy Study Group, *Nuclear Power,* chap. 1.

9. Baruch Fischoff, et al., "How Safe is Safe Enough? A Psychometric Study of Attitudes towards Technological Risks and Benefits," *Policy Sciences* 9 (1978): 127–52.

10. Joel Yellin, "Judicial Review and Nuclear Power: Assessing the Risks of Environmental Catastrophe," *George Washington Law Revue* 45 (1977): 969–93.

11. Daniel Martin, *Three Mile Island: Prologue or Epilogue* (Cambridge, Mass.: Ballinger, 1980); Rufus Miles, *Awakening from the American Dream: The Social and Political Limits to Growth* (New York: Universe, 1976).

12. President's Commission on the Accident at Three Mile Island, *Report of the President's Commission on the Accident at Three Mile Island* (Washington, D.C.: Government Printing Office, 1979); Roger Kasperson, et al., "Institutional Responses to Three Mile Island," *Bulletin of the Atomic Scientists,* 35 (December 1979): 20–24.

13. Robert Goodin, "No Moral Nukes," *Ethics* 90 (1980):417–19; John Elster, "Risk, Uncertainty and Nuclear Power," *Social Science Information* 18 (1979): 371–400.

14. CONAES, *Energy in Transition,* pp. 54 and 459.

15. Lester Lave and Eugene Seskin, *Air Pollution and Human Health* (Baltimore: The Johns Hopkins University Press, 1978); Richard Tobin, *The Social Gamble: Determining Acceptable Levels of Air Quality* (Lexington, Mass.: Lexington, 1979); Richard Wilson and Edmund Crouch, *Risk-Benefit Analysis* (Cambridge, Mass.: Ballinger, 1980), chaps. 2 and 3.

16. *Considerations of Health Benefit-Cost Analysis for Activities Involving Ionizing Radiation Exposure and Alternatives* (National Academy of Sciences, 1977).

17. Lowrance, *Acceptable Risk;* Michael Baram, "Cost-Benefit Analysis: An Inadequate Basis for Health, Safety and Environmental Regulatory Decisionmaking," *Ecology Law Quarterly* 8 (1980):473–532.

18. *Environmental Quality 1979* (Council on Environmental Quality, 1979), p. 657; Gus Spaeth, "A Small Price to Pay," *Environment* 20 (October 1978):25–29.

19. *Environmental Quality 1980* (Council on Environmental Quality, 1980), p. 389.

20. Jeffrey Smith, "EPA and Industry Pursue Regulatory Options," *Science* 211 (1981):796–98.

21. Aaron Wildavsky, "Richer Is Safer," *Public Interest* 60 (Summer 1980): 23–39.

22. *1978 Statistical Yearbook* (New York: United Nations, 1979).

23. Roderick Nash, ed., *The American Environment: Readings in the History of Conservation,* 2nd ed. (Reading, Mass.: Addison-Wesley, 1976); Joseph Petulla, *American Environmentalism: Values, Tactics, Priorities* (College Station, Tex.: Texas A & M University Press, 1980).

24. Ian G. Barbour, ed., *Western Man and Environmental Ethics* (Reading, Mass.: Addison-Wesley, 1973); Thomas Derr, *Ecology and Human Need* (Philadelphia: Westminster Press, 1975).

25. *Public Opinion on Environmental Issues* (Council on Environmental Quality, 1980); *Newsweek,* June 29, 1981, p. 29.

26. *The Direct Use of Coal* (Office of Technology Assessment, 1979); *Energy and the Fate of Ecosystems,* CONAES Supporting Paper No. 8 (National Academy of Sciences, 1980); *Surface Mining: Soil, Coal and Society* (National Academy of Sciences, 1981).

27. Richard Wilson, et al., *Health Effects of Fossil Fuel Burning: Assessment and*

Mitigation (Cambridge, Mass.: Ballinger, 1980); Lave and Seskin, *Air Pollution.*

28. *Energy and Climate* (National Academy of Sciences, 1977); George Woodwell, "The Carbon Dioxide Question," *Scientific American* 238 (January 1978): 34–43. Harold W. Bernard, *The Greenhouse Effect* (Cambridge, Mass.: Ballinger, 1980).

29. Henry Kendall and Steven Nadis, eds., *Energy Strategies: Toward a Solar Future* (Cambridge, Mass.: Ballinger, 1980).

30. John Holdren, Gregory Morris, and Irving Mintzer, "Environmental Aspects of Renewable Energy Sources," *Annual Review of Energy* 5 (1980):241–91.

31. Herbert Inhaber, "Risks with Energy from Conventional and Nonconventional Sources," *Science* 203 (1979):718–23; replies by Rein Lemberg, Richard Caputo, and John Holdren in *Science,* 204 (1979):454, 564–68. See also Chris Whipple, "Energy Impacts of Solar Heating," *Science* 208 (1980):262–66.

32. Henry Hurwitz, "Indoor Air Pollution," *Bulletin of the Atomic Scientists* 37 (February 1981):61–62; replies by Anthony Nero and Jan Bayea, pp. 62–64.

33. Michael Walzer, *Just and Unjust Wars* (New York: Basic Books, 1977), chap. 17; Harold Ford and Francis X. Winters, *Ethics and Nuclear Strategy* (Maryknoll, N.Y.: Orbis, 1977).

34. M. Willrich and T.B. Taylor, *Nuclear Theft;* Nuclear Energy Policy Group, *Nuclear Power;* CONAES, *Energy in Transition.*

35. Gene Rochlin, *Plutonium, Power and Politics* (Berkeley, Calif.: University of California Press, 1979); Stephen Salaff, "The Plutonium Connection: Energy and Arms," *Bulletin of the Atomic Scientists* 36 (September 1980):18–23; Eliot Marshall, "INFCE: Little Progress in Controlling Nuclear Proliferation," *Science* 207 (1980): 1446–47.

36. Harold Feiveson, Frank von Hippel, and Robert Williams, "Fission Power: An Evolutionary Approach," *Science* 203 (1979):330–37; Charles Till, "Technical Considerations in Decisions on Plutonium Use," in *National Energy Issues: How Do We Decide?* ed. Robert Sachs (Cambridge, Mass.: Ballinger, 1980).

37. Frederick Williams and David Deese, eds., *Nuclear Nonproliferation: The Spent Fuel Problem* (New York: Pergamon, 1980).

38. David A. Deese and Joseph S. Nye, *Energy and Security* (Cambridge, Mass.: Ballinger, 1981), chaps. 9–13; Alvin Alm, "Energy Supply Interruptions and National Security," *Science* 211 (1981):1379–84.

39. Samuel Hays, *Conservation and the Gospel of Efficiency: The Progressive Conservation Movement, 1890–1920* (Cambridge, Mass.: Harvard University Press, 1959).

40. See Hessel, *Energy and Ethics* and Abrecht, *Faith and Science.*

41. G. Tyler Miller, *Living in the Environment* (Belmont, Calif.: Wadsworth, 1975).

42. Talbot Page, *Conservation and Economic Efficiency* (Baltimore: The Johns Hopkins University Press, 1977), chap. 7; Derek Parfit, "Energy Policy and the Further Future," working paper, Center for Philosophy and Public Policy, University of Maryland, 1981, to appear in *Energy and The Future,* ed. Douglas MacLean and Peter Brown (Totowa, N.J.: Rowman and Littlefield).

43. R.I. Sikora and Brian Barry, eds., *Obligations to Future Generations* (Philadelphia: Temple University Press, 1978); Peter Singer, "A Utilitarian Popula-

tion Principle," in *Ethics and Population,* ed. Michael Bayles (Cambridge, Mass.: Schenkman, 1976); Ernest Partridge, ed., *Obligations to Future Generations* (Buffalo: Prometheus, 1980).

44. R. Routley and V. Routley, "Nuclear Energy and Obligations to the Future," *Inquiry* 21 (1978):133–79; Ronald Green, "Intergenerational Justice and Environmental Responsibility," *Bioscience* 27 (1977):260–65; D. Clayton Hubin, "Justice and Future Generations," *Philosophy and Public Affairs* 6 (1976):70–83.

45. Mary B. Williams, "Discounting versus Maximum Sustainable Yield," in Sikora and Barry, *Obligations;* Page, *Conservation,* chap. 8.

46. Brian Barry, "Justice Between Generations," in *Law, Morality and Society,* ed. P.M.S. Hacker and J. Raz (Oxford: Clarendon Press, 1977); Barry, "Intergenerational Justice in Energy Policy" working paper, Center for Philosophy and Public Policy, University of Maryland, 1981, to appear in MacLean and Brown, *Energy.*

47. Page, *Conservation,* p. 9.

48. Barry, "Intergenerational Justice."

49. John Holdren, "Fusion Energy in Context," *Science* 20 (1978):168–80; John Clarke, "The Next Step in Fusion: What It Is and How It Is Being Taken," *Science* 210 (1980):967–72.

50. *Application of Solar Technology to Today's Energy Needs* (Office of Technology Assessment, 1978); William Metz and Allen Hammond, *Solar Energy in America* (Washington, D.C.: American Association for the Advancement of Science, 1978).

51. Kendall and Nadis, eds., *Strategies;* John Minan and William Lawrence, eds., *Legal Aspects of Solar Energy* (Lexington, Mass.: Lexington, 1981).

52. Denis Hayes, *Rayes of Hope; Energy from Biomass* (Office of Technology Assessment, 1980).

53. Solar contribution by the year 2000 is estimated as 8 percent by CONAES, 20 percent by Stobaugh and Yergin, 23 percent by CEQ, and 35 percent by the Union of Concerned Scientists (Stobaugh and Yergin, *Energy Future,* p. 263). The Interagency Solar Policy Committee gives 20 percent (*Science* 203 (1979):252).

54. 1978 Battelle Institute study, cited by Stobaugh and Yergin, *Energy Future,* p. 9.

55. Ibid., chap. 7.

56. CONAES, *Energy in Transition,* p. 298. See also Andrew Brook, "Uranium Mine Tailings and Obligations to Future Generations," in *Moral and Ethical Issues Relating to Nuclear Energy Generation* (Toronto, Ontario: Canadian Nuclear Association, 1980).

57. Rustum Roy, "The Technology of Nuclear Waste Management," *Technology Review* 83 (April 1981):39–50; Ronnie Lipschutz, *Radioactive Waste: Politics, Technology and Risk* (Cambridge, Mass.: Ballinger, 1980).

58. Alvin Weinberg, "Social Institutions and Nuclear Energy," *Science* 177 (1972):27–34.

59. Environmental Protection Agency, "Criteria for Radioactive Waste," *Federal Register* 43 (November 15, 1978):53263.

60. Mason Willrich and Richard Lester, *Radioactive Waste: Management and Regulation* (New York: The Free Press, 1977); "Report to the American Physical Society by the Study Group on Nuclear Fuel Cycles and Waste Management," *Review of Modern Physics* 50 (1980):1–186.

61. "Panel Throws Doubt on Vitrification," *Science* 201 (1978):599; Interagency Review Group on Nuclear Waste Management, *Report to the President* (U.S. Department of Energy, 1979).

62. Roger Kasperson, "The Dark Side of the Radioactive Waste Problem," in *Progress in Resource and Environmental Planning,* ed. T. O'Riordan and K. Turner (New York: John Wiley & Sons, 1980); Gene Rochlin, "Nuclear Waste Disposal: Two Social Criteria," *Science* 195 (1977):23–31; Kai Lee, "A Federalist Strategy for Nuclear Waste Management," *Science* 208 (1980):679–84.

CHAPTER 6: VALUES AND POLICIES: SOCIOPOLITICAL DILEMMAS

1. Report of the NAACP National Energy Conference (Washington, D.C., NAACP, December 21, 1977).

2. William Winpisinger, a trade union leader, has denounced nuclear power because it "kills workers through radiation" and forces them to work in "cancerous cesspools." His remarks were reported in the *Los Angeles Times,* April 12, 1981.

3. H. Alfven, *Bulletin of the Atomic Scientists* 28 (May 1972):5; Amory B. Lovins, *World Energy Strategies* (Cambridge, Mass.: Ballinger, 1975), p. 62.

4. W.M. Hawkins, "Commercial Aircraft Experience," in *The Outlook for Nuclear Power,* presentations at the Technical Sessions, National Academy of Engineering, November 1, 1979, pp. 39–44.

5. For a summary of existing estimates see W. Häfele, et al., *Energy in a Finite World* (Cambridge, Mass.: Ballinger, 1981), esp. pp. 122–28 and table 7–5.

6. Center for Philosophy and Public Policy, "Why People Fear Nuclear Power," report from the Center based on article by D. MacLean; Nelkin and Fallows, "The Evolution of the Nuclear Debate."

7. L. Nader, "The Politics of Energy: Toward a Bottom-Up Approach," *Radcliffe Quarterly* 67 (December 1981): 5–6.

8. W. Häfele, et al., *Energy in a Finite World,* chap. 6 and chap. 7, pp. 100–4.

9. D.F. Ford, H.W. Kendall, *An Assessment of the Emergency Core Cooling Systems Rulemaking Hearings* (San Francisco: Friends of the Earth, 1974).

10. Cf., for example, Louis Harris and Associates, *A Survey of Public and Leadership Attitudes Towards Nuclear Power Development in the United States* (New York: Ebasco Services, 1975).

11. It can be argued, however, that the experience of Hiroshima and Nagasaki did not prove a negative. It was not sufficiently sensitive to exclude effects that might be significant on a worldwide scale after large-scale deployment of nuclear technology.

12. R.E. Kasperson, G. Berk, D. Pijawka, A.B. Sharaf, and J. Wood, "Public Opposition to Nuclear Power: Retrospect and Prospect," in *Supporting Paper 5: Sociopolitical Effects of Energy Use and Policy,* Committee on Nuclear and Alternative Energy Systems (Washington, D.C.: National Academy of Sciences, 1979), pp. 259–92.

13. "The American Energy Consumer: Rich, Poor, and in Between," in *A Time to Choose* (Cambridge, Mass.: Ballinger, 1974), pp. 113–30.

14. P.K. Verleger, Jr., "Alternatives Available to Reduce Impacts of a Supply Interruption on Transportation—Lessons from the Crisis of the 70's," Harvard

University Energy and Environmental Policy Center paper E-80-10, Cambridge, Mass., (December 1980).

15. Alm, Colglazier, and Kates-Garnick, "Coping With Interruptions," in Deese and Nye, eds., *Energy Security.*

16. T. Neville, "What Price Electricity?" draft paper for South West Energy Policy Study (SWEPS), Graduate School of Business, Stanford University, March 17, 1976; E.R. Habicht, *The Natural Gas Sector: Rational Fuel Conversion Strategies, Pricing Policies and Investment Instruments* (Environmental Defense Fund, February 11, 1977).

17. C. Wolf, Jr., "A Theory of Non-Market Failures," *The Public Interest* 55 (Spring 1979): 114–33.

18. D. Usher, *The Economic Prerequisite to Democracy* (New York: Columbia University Press, 1981).

19. P.G. Peterson, "No More Free Lunch for the Middle Class," *The New York Times Sunday Magazine,* January 17, 1982.

20. G. Stigler, "The Economics of Minimum Wage Legislation," *American Economic Review,* 36 (June 1946): 358; M. Friedman, *Capitalism and Freedom* (Chicago: University of Chicago Press, 1962); cf. also H. Heclo and M. Rein, "Social Science and Negative Income Taxation," in OECD, *The Utilization of the Social Sciences in Policy Making in the United States* (Paris: Organization for Economic Cooperation and Development, 1980), pp. 29–66.

21. *Wall Street Journal,* October 26, 1979.

22. *The New York Times,* November 30, 1980, p. F2.

23. For a detailed proposal of this type cf. L.M. Greene, *Free Enterprise Without Poverty* (New York, W.W. Norton, 1981).

24. C. Schultze, *The Public Use of Private Interest* (Washington, D.C.: Brookings Institution, 1977); F. Anderson, et al., *Environmental Improvement Through Economic Incentives* (Baltimore: The Johns Hopkins University Press, 1977).

25. R.A. Liroff, *Air Pollution Offsets: Trading, Selling, and Banking, An Issue Report* (Washington, D.C.: The Conservation Foundation, 1981); A. Rosencranz, "Economic Approaches to Air Pollution Control," *Environment,* 23 (October 1981): 25–30.

26. Meadows, et al., *The Limits to Growth;* A. Peccei, *The Chasm Ahead* (London: MacMillan, 1969). For a comprehensive critical review of the debate on world system models see M. Greenberger, M.A. Crenson, and B.L. Crissey, *Models in the Policy Process: Public Decision Making in the Computer Era* (New York: Russell Sage Foundation, 1976), pp. 158–82.

27. N. Keyfitz, "World Resources and the World Middle Class," *Scientific American* (July 1976).

28. H. Brooks, "Technology: Hope or Catastrophe?", *Technology in Society* 1 (1979): 3–17.

29. H. Brooks, "Can Technology Assure Unending Material Progress?", in *Progress and Its Discontents,* ed. G. Almond, M. Chodorow, and R.H. Pearce (Berkeley, Calif.: University of California Press, 1982).

30. L.A. Orleans, "Science, Elitism, and Economic Readjustment in China," *Science,* 215 (January 29, 1982):472–77.

31. Usher, *Economic Prerequisite,* chap. 6.

32. J. Gibbons, et al., *Alternative Energy Demand Future to 2010,* Report of the Demand and Conservation Panel, Committee on Nuclear and Alternative Energy Systems (Washington, D.C.: National Academy of Sciences, 1979), pp. 111–17.

33. Cf., for example, R.M. Parsons Co., *Preliminary Design Study for An Integrated Coal Gasification Combined Cycle Power Plant* (Electric Power Research Institute, 1978).

34. H. Brooks, "A Critique of the Concept of Appropriate Technology," in *Appropriate Technology and Social Values: A Critical Appraisal,* ed. F.A. Long and A. Oleson (Cambridge, Mass.: Ballinger, 1980).

35. W. Häfele, et al., *Energy in a Finite World,* chap. 6.

36. M. Simmons, et al., *Supporting Paper 6: Domestic Potential of Solar and Other Renewable Energy Sources,* Committee on Nuclear and Alternative Energy Systems (Washington, D.C.: National Academy of Sciences, 1979).

37. W. Häfele, et al., *Energy in a Finite World,* chap. 7.

CHAPTER 7: VALUES AND POLICIES: DILEMMAS IN RISK ASSESSMENT

1. W. Häfele, et al., *Energy in a Finite World,* p. 116, figure 7–9.

2. National Research Council, *Energy in Transition 1985–2010,* Committee on Nuclear and Alternative Energy Systems (San Francisco: W.H. Freeman, 1979), p. 476; J. Harte, et al., *Supporting Paper 8: Energy and the Fate of Ecosystems* Committee on Nuclear and Alternative Energy Systems (Washington, D.C.: National Academy of Sciences, 1980).

3. W. Häfele, et al., *Energy in a Finite World,* chap. 7, pp. 122–27, esp. table 7–5.

4. Cf., for example, "Radiation Exposures of Hanford Workers: A Critique of the Mancuso, Stewart, and Kneale Report," *Health Physics,* 34 (1978): 744–50.

5. Office of Technology Assessment (OTA), *Assessment of Technologies for Determining Cancer Risks to the Environment* (Washington, D.C.: U.S. Government Printing Office, June 1981).

6. A. Wildavsky, "Richer Is Safer," *The Public Interest* 60 (Summer, 1980): 23–39.

7. S. Black, F. Niehaus, and D. Simpson, *How Safe is Too Safe?,* WP-79-68 (Laxenburg, Austria: International Institute for Applied Systems Analysis, 1979).

8. A.O. Hirschman, "The Principle of the Hiding Hand," *The Public Interest* 6 (Winter 1967): 10–23.

9. J. Crow, et al., *Risks and Impacts of Alternative Energy Systems,* Risk and Impact Panel, Committee on Nuclear and Alternative Energy Systems, Washington, D.C., National Academy of Sciences, to be published.

10. R. Revelle, "Let the Waters Bring Forth Abundantly," in *Arid Zone Development: Potentialities and Problems* ed. Y. Mundlak, S.F. Singer (Cambridge, Mass.: Ballinger, 1977), chap. 14, p. 191.

11. C. Starr, "Benefit Cost Studies in Sociotechnical Systems," in *Perspectives in Benefit-Risk Decision Making* (Washington, D.C.: National Academy of Engineering, 1971).

12. Office of Science and Technology (OST), *Cumulative Regulatory Effects on the Cost of Automotive Transportation, RECAT* (Washington, D.C.: U.S. Government Printing Office, February 28, 1972).

13. N. Rasmussen, et al., *Reactor Safety Study, WASH 1400 or NUREG-75-014* (Washington, D.C.: U.S. Nuclear Regulatory Commission, 1975); cf. also, H.W. Lewis, et al., *Report of the Risk Assessment Group to the U.S. Nuclear Regulatory Commission, NUREG/CR-0400* (Washington, D.C.: U.S. Government Printing Office, 1978).

14. CONAES, *Energy in Transition,* pp. 284–89.

15. Hurwitz, "Indoor Air Pollution" and replies by Nero and Bayea.

16. CONAES, *Energy in Transition,* chap. 5.

CHAPTER 8: ENERGY AND DEMOCRACY: VALUES AND PROCESS

1. Joel Darmstadter, Joy Dunkerley, and Jack Alterman, "International variations in Energy Use: Findings from a Comparative Study," *Annual Review of Energy* 3 (1978):203.

2. Harvey Brooks and Jack M. Hollander, "United States Energy Alternatives to 2010 and Beyond: The CONAES Study," *Annual Review of Energy* 4 (1979):4–5.

3. National Geographic, *Energy* (Washington, D.C.: National Geographic Society, February 1981), p. 36.

4. Within the consensual framework, there is also considerable disagreement. Americans have often divided into factions and parties, into interest groups with conflicting concerns and beliefs. These divisions have led to passionate disagreements and to struggles over power and public policy. To conservatives, the democratic ideal means that government must be kept in check so that as many economic decisions as possible will be made by individuals and private groups—even if the result is that market forces determine the allocation and distribution of most resources other than those necessary for collective security. To liberals, true democracy means that the state must regulate the economy and intervene in it to insure economic security and welfare and to expand opportunities as widely as possible. Many, though not all, conservatives also believe that democracy authorizes the majority to assure that certain moral standards are observed. Virtually all liberals, and some "libertarian" conservatives, believe that only the individual person has the right to impose moral standards upon himself, unless his behavior is harmful to others. American political argument reflects not only the clash of economic and regional interests but also a clash over these beliefs concerning the meaning of democracy.

5. For a view that inequality has been significantly diminished see Edgar K. Browning, "How Much More Equality Can We Afford?", *The Public Interest* 43 (Spring 1976): 90–110. For a critique of this view see Timothy Smedding, "The Trend Toward Equality in the Distribution of Net Income," Wisconsin Institute for Research on Poverty (December 1977).

6. Lester C. Thurow, *The Zero-Sum Society* (New York: Basic Books, 1980).

7. See Stanley Rothman and S. Robert Lichter, "Power, Politics, and Personality in Post-Industrial Society," *Journal of Politics,* 40 (August 1978): 675–707.

8. It has been pointed out that auto registration records, on which the issuance of ration cards would have to be based, are not completely accurate; that coupon counterfeiting would be hard to prevent; that the need to make exceptions would require the creation of a significant new bureaucracy at the local level; and that supply would not match demand in every region, at least in the early stages. A better alternative might be to decontrol prices and impose an emergency windfall profits tax on markups over the price of crude oil, which would be related to owners of automobiles or to households. See Alm, Coleglazier, Kates–Garnick, "Coping with Interruptions," in Deese and Nye, eds, *Energy and Security,* pp. 341–44.

9. Robert C. Wood, *1400 Governments: The Political Economy of the New York Metropolitan Region* (Garden City, N.Y.: Doubleday, 1964).

10. See Melvin M. Webber, "The BART Experience: What Have We Learned?," *The Public Interest* 45 (Fall 1976):79–108.

11. See Robert A. Dahl, *Polyarchy: Participation and Opposition* (New Haven, Conn.: Yale University Press, 1971).

12. Nuclear Energy Policy Study Group (the Ford-Mitre Report), *Nuclear Power Issues and Choices* (Cambridge, Mass.: Ballinger, 1977), p. 159.

13. See Charles O. Jones, "American Politics and the Organization of Energy Decision Making," *Annual Review of Energy* 4 (1979):99–121.

14. Crauford D. Goodwin, "Conclusion," in Crauford D. Goodwin, editor, *Energy Policy in Perspective* (Washington, D.C.: Brookings Institute, 1981), p. 673. See also Richard B. Mancke, *The Failure of U.S. Energy Policy* (New York: Columbia University Press, 1974), p. 143.

15. Goodwin, *Energy Policy,* p. 681.

16. Dorothy Nelkin and Susan Fallows, "The Evolution of the Nuclear Debate," *Ann. Rev. Energy* 3 (1978), p. 277.

17. Ibid., p. 279.

18. Ibid., p. 284.

19. Ibid., pp. 290–91.

20. Ibid., p. 299.

21. See S.A. Lakoff, "Congress and National Science Policy," *Political Science Quarterly* 89 (Fall 1974):589–611.

22. C.P. Snow, *Science and Government* (Cambridge, Mass.: Harvard University Press, 1961).

23. See S.A. Lakoff, "Scientists, Technologists and Political Power," in *Science, Technology and Society* and "Ethical Responsibility and the Scientific Vocation," S.A. Lakoff, ed. *Science and Ethical Responsibility* (Reading, Mass.: Addison-Wesley, 1980).

24. For a review of the literature treating the recombinant DNA controversy and the efforts to resolve it by combining expert judgment and popular participation, see S.A. Lakoff, "Moral Responsibility and the 'Galilean Imperative,'" *Ethics* 91 (October 1980):100–16.

25. This discussion draws on the account of the experiment in Milton R. Wessel, *Science and Conscience* (New York: Columbia University Press, 1980), pp. 165–79.

CHAPTER 9: AMERICA AND THE WORLD

1. Energy Systems Program Group, International Institute for Applied Systems Analysis, *Energy in a Finite World: Paths to a Sustainable Future* (Cambridge, Mass.: Ballinger, 1981).
2. See Joseph S. Nye, "Energy and Security," Deese and Nye, eds., *Energy and Security,* pp. 15–16.
3. N.J.D. Lucas, *Energy and the European Communities* (London: Europa Publications, 1977), p. 154.
4. Gerard Smith and George Rathjens, "Reassessing Nuclear Nonproliferation Policy," *Foreign Affairs* 59 (Spring 1981):888.
5. Ibid., 890.
6. For an exposition and defense of the proposal see Jan Tinbergen, *RIO: Reshaping the International Order* (New York: Dutton, 1976).
7. See Robert Tucker, "Oil: The Issue of American Intervention," *Commentary* 59 (January 1975): 21–31.
8. Roger Revelle, "Energy and Development: The Case of Asia," Lakoff, ed., *Science and Ethical Responsibility* pp. 299–300.
9. Ibid., p. 291.
10. David A. Deese, "The Oil-Importing Developing Countries," Deese and Nye, eds., *Energy and Security,* pp. 230–31.
11. Ibid., p. 245.
12. Revelle, "Energy and Development," in Lakoff, ed., *Science,* p. 299.
13. *Global Two Thousand Report to the President of the United States,* edited by Gerald O. Barney (New York: Pergamon, 1980) vol. 1, p. 17.
14. Ibid., p. 2.
15. Deese, "Oil-Importing" in Deese and Nye, eds., *Energy and Security,* p. 233.
16. As reported in *The New York Times,* June 5, 1981, p. 26.
17. Jahangir Amuzegar calculates that the annual OPEC gain amounted to 1.7 percent of the "rich North's" GNP rather than the 2 percent intended, and less if purchases of LDC imports are subtracted. As he also points out, "almost the entire burden of payment imbalances caused by the oil price rises was shifted to the poor and helpless LDCs. For the whole 1973–79 period, the North had a combined surplus of $32.5 billion, while nonoil LDCs incurred a whopping total deficit of some $245 billion. OPEC's surpluses were thus almost totally borrowed by the developing world through official and private recycling." Amuzegar, "Petrodollars Again," *The Washington Quarterly* 4 (Winter 1981):137–38.
18. According to Leonard Silk, *The New York Times,* June 5, 1981, p. 26.
19. Verleger, Jr., "Financial Implications" in Deese and Nye, eds., *Energy and Security,* p. 347.
20. Deese, "Oil-Importing," in Deese and Nye, eds., *Energy and Security,* p. 243.

CHAPTER 10: PRESENT NEEDS AND FUTURE PROSPECTS

1. Thomas Widmer and Elias Gyftopoulos, "Energy Conservation and a Healthy Economy," *Technology Review* 79 (June 1977):33.

2. Stobaugh and Yergin, eds., *Energy Future,* p. 229.

3. Studies by the Mellon Institute, National Audubon Society, and Solar Energy Research Institute are summarized in Norman Colin, "Energy Conservation: The Debate Begins," *Science* 212 (1981):424–26.

4. Robert H. Williams, ed., *The Energy Conservation Papers* (Cambridge, Mass.: Ballinger, 1975), chaps. 2, 3; Eric Hirst, "Transportation Energy Conservation Policies," *Science* 192 (1976):15–20; Gerald Leach, *A Low Energy Strategy for the United Kingdom* (London: International Institute for Environment and Development, 1979).

5. Marc Ross and Robert Williams, *Our Energy: Regaining Control* (New York: McGraw-Hill, 1981); National Energy Strategies Project (Sam Schurr, director), *Energy in America's Future* (Baltimore: The Johns Hopkins University Press, 1979).

6. Eric Hirst and Janet Carney, "Effects of Residential Energy Conservation Programs," *Science* 199 (1978):845–51; William Brownell and Loren Dunn, *Federal Energy Conservation Policy* (Lexington, Mass.: Lexington, 1980).

7. Seymour Warkov, ed., *Energy Policy in the U.S.: Social and Behavioral Dimensions* (New York: Praeger, 1978); "Motivating the Troops for the Energy War," *Psychology Today,* April, 1979, pp. 14–33; Paul Stern and Eileen Kirkpatrick, "Energy Behavior," *Environment* 19 (December 1977):10–15; Robert Socolow, ed., *Saving Energy in the Home: Princeton's Experiments at Twin Rivers* (Cambridge, Mass.: Ballinger, 1979).

8. Lee Schipper and Joel Darmstadter, "The Logic of Energy Conservation," *Technology Review* 80 (January 1978):46.

9. Maxey, "Nuclear Energy Politics"; Rossin, "The Soft Energy Path" and "Centralization."

10. William Leiss, *The Limits of Satisfaction* (Toronto: University of Toronto Press, 1976).

11. Paul Abrecht, *Faith and Science;* John Taylor, *Enough is Enough* (Minneapolis: Augsburg, 1977); Larry Rasmussen and Bruce Birch, *The Predicament of the Prosperous* (Philadelphia: Westminster, 1978).

12. On energy and lifestyles, see chapter by C.P. Wolf in *Sociopolitical Effects of Energy Use and Policy,* CONAES Supporting Paper No. 5 (National Academy of Sciences, 1979); also chap. 5 of *Energy Choices in a Democratic Society,* CONAES Supporting Paper No. 7 (National Academy of Sciences, 1980); Komon Valaskakis, et al., *The Conserver Society* (New York: Harper & Row, 1979).

13. M. Lönnroth, T.B. Johansson, and P. Steen, "Sweden Beyond Oil: Nuclear Commitments and Solar Options," *Science* 208 (1980):557–63.

14. *Public Opinion on Environmental Issues* (Council on Environmental Quality, 1980), pp. 22, 23.

15. Robert Mitchell, "The Public Response to a Highly Publicized Major Failure of a Controversial Technology" (Washington, D.C.: Resources for the Future, Discussion Paper D-60, 1980).

16. Robert Kasperson, Robert Kates, Mundo Morrison, and Barry Rubin, "Value Issues in the Nuclear Power Controversy," in *The Closure of Scientific Debates,* ed. H. Tristram Englehardt and Arthur Caplan (New York: Plenum, 1982).

17. National Energy Strategies Project; Ramsey, *Unpaid Costs,* chap. 20.

Index

About the Authors

IAN BARBOUR is Winifred and Atherton Bean Professor of Science, Technology, and Society at Carleton College, Northfield, Minnesota. He received his Ph.D. degree in physics from the University of Chicago and the B.D. degree in theology from Yale. He has been awarded Guggenheim and Fulbright fellowships. Among his writings are *Issues in Science and Religion* and *Technology, Environment and Human Values* (Praeger).

HARVEY BROOKS is Benjamin Pierce Professor of Technology and Public Policy at Harvard, where he received his doctorate in physics and where he has also served as Gordon McKay Professor of Applied Physics. He has served on many high-level commissions and was cochairman of the National Academy of Sciences Committee on Nuclear and Alternative Energy Sources. The author of *The Government of Science* and of many scientific papers and articles on science policy, he has received numerous awards, including several honorary degrees as well as the E.O. Lawrence Award of the Atomic Energy Commission and the Kazan Medal from Columbia University.

SANFORD LAKOFF is Professor of Political Science and an affiliate of the Program in Science Technology and Public Affairs at the University of California, San Diego. He received his doctorate from Harvard and is the author of *Equality in Political Philosophy,* coauthor of *Science and the Nation,* and editor of several other books, including *Science and Ethical Responsibility.*

JOHN OPIE, who received his doctorate from the University of Chicago and a B.D. degree from Union Theological Seminary, is Professor of History at Duquesne University. The author of *Americans and Environment: The Controversy Over Ecology* and *Jonathan Edwards and the Enlightenement,* he is also founding editor of *Environmental Review,* an interdisciplinary journal.